T0305596

Digital Project Practice for New Work and Industry 4.0

New Work and Industry 4.0 have matured, and this book takes a practical, experience-based approach to project management in these areas. It introduces methods and covers the practical aspects. It critically examines existing approaches and practices and shows their limitations. The book covers appropriate methods as well as human and social aspects. It contributes to the ongoing discussion of business practices and methods. It also aims to stimulate dialogue in the professional community.

Digital Project Practice for New Work and Industry 4.0 begins by introducing basic concepts in the context of Industry 4.0 and discussing how they might influence organizational communication and impact the work environment. After examining the possibilities and challenges of remote work and collaboration in distributed teams all over the world, the book looks at a company's fundamental changes related to New Work from a practical business perspective as well as legal and ethical perspectives. It reviews the case of the VW emission scandal and recommends ways to improve corporate culture. Legal issues include New Work and hybrid forms of collaboration as well as liability for automated decisions (i.e., the potential need for an "electronic person"). Other implications for the workplace include how:

- Industry 4.0 might influence the potential demand for "Digital Unions"
- Industry 4.0, lean production, and their applications can change industrial practices
- Open Banking presents new approaches and new business models
- Work structures and systems can empower employees' work self-management

This book also looks at how New Work affects individual workers. It addresses digital stress, introduces strategies for coping with it, and discusses related topics. It also explores the benefits of meditation and the economics of mind, body, and spirit.

In essence, this book covers appropriate methods along with human and social factors. It also covers practice, different perspectives, and various experiences from all around the globe. Contributing to the ongoing discussion on business practices and methods, this book will nourish and stimulate dialogue in the professional community.

Digital Project Practice for New Work and Industry 4.0

Edited by
Tobias Endress

CRC Press
Taylor & Francis Group
Boca Raton London New York

CRC Press is an imprint of the
Taylor & Francis Group, an **informa** business
AN AUERBACH BOOK

Cover image: Olena Shkavron

First edition published 2023
by CRC Press
6000 Broken Sound Parkway NW, Suite 300, Boca Raton, FL 33487-2742

and by CRC Press
4 Park Square, Milton Park, Abingdon, Oxon, OX14 4RN

CRC Press is an imprint of Taylor & Francis Group, LLC

© 2023 Taylor & Francis Group, LLC

ISBN: 9781032438986 (hbk)
ISBN: 9781032276045 (pbk)
ISBN: 9781003371397 (ebk)

DOI: 10.1201/9781003371397

Typeset in Minion
by Newgen Publishing UK

Contents

Preface

Dear Reader,

New Work and Industry 4.0 has come to maturity, and comprehensive reports with the practical experience from the field can be compiled. "Always-online" business environments, Industry 4.0 strategies and New Work have become the new normal and significantly impact our daily routines. The various chapters in this book provide a rich selection of topics, experiences, and opinions from experts in the field.

Chapter 1 introduces some basic concepts in the context of Industry 4.0 and discusses how they might influence our communication and impact our work environment. Chapters 2 and 3 address the given possibilities and challenges with remote work and collaboration in distributed teams all over the world. Chapter 4 presents a company's fundamental changes related to New Work from a practical business perspective. In Chapter 5, we review the case of the VW emission scandal and what we can learn from it in terms of the improved corporate culture. Chapter 6 advances further on how Industry 4.0 might influence our work environment and also explores aspects like the potential demand for "Digital Unions." Chapter 7 addresses digital stress and introduces strategies for coping with it, while Chapter 8 discusses related topics from a different perspective. It propounds to consider meditation and economics of mind, body, and spirit. Chapters 9 and 10 cover several legal issues, including New Work and hybrid forms of collaboration as well as liability for automated decisions (i.e., the potential need for an "electronic person"). Chapter 11 reviews Industry 4.0 and lean production (LP) and their applications in industrial practices. Chapter 12 discusses the concrete application of Open Banking approaches and the potential impact of this transformation, including the new business models and opportunities that arise from it for the financial industry. Finally, Chapter 13 takes up new self-managing work systems and how organizations should create work structure and system that empowers their employees' work self-management. Although the coverage of the topic in this volume can, of course, by no means be complete. I think that this selection gives a very good overview and covers a variety of interesting perspectives.

The book at hand is the 2nd volume of the "Digital Project Practice" book series. The previous book, *Digital Project Practice: Managing Innovation &*

Change, received excellent feedback and reviews. It has become a valuable resource for practitioners as well as academics. This encouraged the co-authors and me to start working on the 2nd volume. The book series introduces the relevant methods, covers the practical aspects, critically acclaims existing approaches and practices, and shows limitations. In essence, the books cover appropriate methods along with human/social factors. In fact, the human element is one of the running themes in the book series. You can regard it as a contribution to the ongoing business practices and methods discussion. It also aims to nourish and stimulate dialogue in the professional community.

In crux, this book is about practice, different perspectives, and various experiences from all around the globe. I hope you enjoy reading it and that it can help you reflect on your organization's current practices.

With best regards,
Dr. Tobias Endress
Lautoka / Fiji Islands, July 5, 2022

Acknowledgments

Illustrations: Olena Shkavron (cover image and figures 1.2, 2.1, 4.1, 5.1, 6.1, 8.1, 9.1, and 12.1), Subrata Dutta/iRiqa Services (figures 1.1, 3.1, 7.1, 10.1, and 11.1)

Proofreader: Umar Afridi, Alyssa Myers

Translation: Dr. Tobias Endress (Chapter 4: Why New Work is the Result of Fundamental Changes in a Company)

Editor Bio

Dr. Tobias Endress is an assistant professor at the Asian Institute of Technology (AIT) | School of Management. He has more than 20 years of professional experience in digital project business and innovation management. His former (non-academic) roles include project manager, product owner and business analyst.

He has completed professional training in banking and graduated in Computer Science and Business Administration at VWA Frankfurt/Main and in Business Economics at Avans+ in Breda (NL). He obtained a Master's Degree in Leadership in Digital Communication at Berlin University of the Arts (UdK Berlin) and a Doctorate in Business Administration at the University of Gloucestershire in Cheltenham (UK). He is a fellow of the EuroMed Academy of Business.

Contributors

Christopher M. Castille is an assistant professor of Management in the Department of Management and Marketing at Nicholls State University. His Ph.D. in Industrial-Organizational Psychology was earned from Louisiana Tech University. At Nicholls, Dr. Castille teaches coursework on Human Relations and Interpersonal Skills, Performance and Compensation Management, HR Analytics, and Managing Human Capital. He also studies the role of personality and individual differences at work, behavioral ethics, and research methods. His work has been featured in academic outlets including *Journal of Business and Psychology*, *Journal of Business Ethics*, *Personality and Individual Differences*, and *Industrial-Organizational Psychology: Perspectives on Science and Practice*.

Kourosh Dadgar is an associate professor of Information Systems and Business Analytics at the School of Management, University of San Francisco. He has published papers in prestigious journals of JAIS (*Journal of Association for Information Systems*), IJIM (*International Journal of Information Management*), JBE (*Journal of Business Ethics*) and presented at the flagship IS conferences of ICIS (International Conference on Information Systems) and HICSS (The Hawaii International Conference on System Sciences). He holds a health IT minitrack at the HICSS conference. His main research areas are ICT-enabled self-management, data mining, and value-sensitive healthcare systems and technologies.

Tracy Dathe is a professor of Business Management at Macromedia University on Campus Berlin, Germany. She is also an experienced finance professional. As CFO of a multinational medium-sized automotive supplier, she was responsible for the overall commercial management at the German headquarters and the overseas subsidiaries in China, France, Italy, Sweden, the Czech Republic, Turkey, and the USA. Today, she also supports medium-sized companies as a management consultant.

David Galipeau is a 35-year veteran of a hybrid career within finance, publishing, digital technology, and social development and has worked in Canada, Europe, and Asia as a corporate executive, dot-com entrepreneur, and social innovator. After completing his graduate education in the USA, he joined the United Nations in 2005, leading several cutting-edge innovation initiatives in Europe, Africa, and Asia-Pacific. Mr. Galipeau left the United Nations in 2019 and co-founded SDGx, an impact investment and advanced technology research organization based in Singapore, Australia, and Germany. Mr. Galipeau also lectures at the United Nations System Staff College, Asian Institute of Technology (AIT), and other universities.

Christiane Hagmann-Steinbach is a consultant with a focus on innovation management and sustainability. Additionally, she is division manager in a welfare association and a lecturer at the Baden-Wuerttemberg Cooperative State University. She has more than 30 years of professional experience.

She obtained a Master's Degree in Leadership in Digital Communication at Berlin University of the Arts, as well as a Master of Economics at the University of St. Gallen.

Marc Helmold is a full-time professor of General Business Administration, Business Ethics, Performance Management and International Negotiations at the IU International University of Applied Sciences at the Berlin campus. Before that he held various managerial positions and was managing director of leading manufacturers in the automotive and rail industries. With this in mind, he was able to expand CSR activities sustainably. In the industry, in particular, he was able to transfer sustainability concepts to the entire value chain. He spent several years in Japan and China.

Günter Jeschke works as a consultant at 1st solution consulting GmbH. He has over 20 years of professional experience in order management processes inside ITK short- and long-term projects. His former roles include project support manager (media events), delivery manager, and consultant for new business development. He has completed professional training for telecommunication electronics and has graduated in Computer Science and Business Administration at VWA Frankfurt am Main, in Business Economics at Avans+ in Breda (NL).

Syed Shurid Khan is a former corporate banker-turned-professor cum personal wellness enthusiast. He is currently working as an assistant professor of economics at the Asian Institute of Technology (AIT) in Bangkok. His research experiences come from working on renewable energy policies of the United States during his time as a researcher with the UHERO – Hawaii's top economic think tank. His research expertise also lies in local food system modeling, financial markets, and applied econometrics. He also explores different areas within behavioral economics, particularly how soft skills (e.g., hobbies, meditation, etc.) rewire our brain and mind and how they influence the notion of personal liberty as well as professional integrity. During leisure time, Syed likes to travel and meditate.

Philipp Lehmkuhl has been driving the digital transformation at the international company MEWA Textil-Management since 2017. After his studies in "Leadership in Digital Communication" in Berlin and St. Gallen, he spent more than 16 years in digital leadership in digital management positions in medium-sized companies and multinational corporations, including as a director and authorized signatory at ALDI Süd, where he led the Customer Interaction department in the International Transformation Organization and International IT. Prior to that, he worked for T-Systems Multimedia Solutions as director of a business unit, built up a foreign branch in China for a management consulting firm and gained experience with his own startup.

Stephan Meyer designs the society of tomorrow. He likes to question "sacred cows" in order to uncover the essentials behind them. He studied business psychology at RWTH Aachen University. He also holds a Ph.D. in Business Administration from the University of Gloucestershire for his research on how to slaughter sacred cows in organizations. Some clients, therefore, call him the "Sacred Cow Exterminator." Dr. Stephan Meyer is active as a trusted advisor, coach, and keynote speaker. A native of Germany, he lives in Cyprus, in an orange grove.

Marc Nathmann is legal counsel and attorney at law at ING Germany in Frankfurt. There he advises on banking supervisory and capital markets law in the areas of payment transactions and securities.

His research interests include FinTech and AI. He regularly gives lectures on banking and capital markets law and commercial law. He has published articles on topics dealing with the legal classification of innovations and new technologies.

After his studies at the University of Münster and the German University of Administrative Sciences in Speyer, he stayed abroad at a major international law firm in Shanghai, China, worked at KPMG, a securities bank and a FinTech consulting firm. He completed his doctorate on the subject of FinTech at the Technical University of Chemnitz.

Martin Schneider is responsible for building up the API Banking activities at Helaba, a leading German commercial bank. Prior to that role, he was heading Commerzbank's CI/CD engineers. Martin is a young leader, passionate about innovation, digitization and cultural change, project management (agile, classic) expertise, and professional experience in the banking and insurance industry. His holds is a Ph.D. (Dr. rer. pol.) in Finance from the University of Potsdam and a Master's degree from the London School of Economics and Political Science, complemented by international work experience in New York, London, Luxembourg, and Moscow.

1

Industry 4.0 – The Impact on Communication and Work Environment

Günter Jeschke
Consultant, 1st solution consulting gmbh, Düsseldorf, Germany

Tobias Endress
Asian Institute of Technology (AIT), Bangkok, Thailand

CONTENTS

BUSINESS STRATEGIES CHANGE DRAMATICALLY THROUGH INDUSTRY 4.0

In the last couple of years, the term "Industry 4.0," or the fourth Industrial Revolution (IR), has been coined to characterize several concepts and strategies, mostly related to the integration of Information Technology (IT) and

DOI: 10.1201/9781003371397-1

Communication (ICT) in production processes (e.g., Internet of Things [IoT], implementation of digital production, and support technologies). Still, it is also about changing business environments, cooperation, artificial intelligence (AI), and a potential need for process alterations. Insofar, Industry 4.0 has been described as more than just implementing new technology. It is also a "cultural philosophy about how we can use increased visibility, flexibility, and efficiency across our production to be more competitive" (Healy, 2018). Klaus Schwab, the World Economic Forum (WEF) founder, pointed out various societal concerns associated with the IR and named inequality the most important one. Still, he also highlighted problems in the context of privacy and new communication patterns. He expressed his concern that "constant connection may deprive us of one of life's most important assets: the time to pause, reflect and engage in meaningful conversation" (Schwab, 2016).

Various research has already been conducted to identify the new challenges and adoption of these transformative practices by companies that have already implemented Industry 4.0 processes (Gupta & Goyal, 2021). We can observe the consequences of the digital transformation within various industries which increase and accelerate the general agenda for production automation. For organizations, it becomes more and more evident that the requirements of competitive production in terms of speed, quality, flexibility, and innovative and individualized products increase significantly. Globalization changes the significance of networking between the organizations massively. Businesses in all industries worldwide are transforming themselves to become more agile, customer friendly, and introduce new business models.

The 4th IR fosters hybridization (linking products and services) and integration of customers and partners into value-added business processes (Bendel, n.d.). However, it is not only about technology and operations. It is about the people involved as well. The "always-on" culture has various impacts on the way we work and interact. It appears to change the working world and the society as a whole—even older people are now affected by this change in culture (Lüders & Gjevjon, 2017). Strategies and concepts like the Internet of People (IoP), IoT, Internet of People, Things & Services (IoPTS), AI, and also several more are becoming integral part of the standard toolbox for every manager.

A comprehensive understanding of digital technologies is no longer exclusively assigned to IT but requires a solid foundation in the organization as a whole. Therefore, digital competence becomes a requirement for many roles, particularly for management and strategy-related tasks. Some authors state that the waves of 5G, AI, and IoT will pave the way to generate new revenue streams but also emphasize the importance of keeping up or even gaining thought leadership in these new markets (Chhaya, 2020). For example, some leading industrial companies, like ABB, clearly state that their strategy is to realign manufacturing around digital industries (Vavra, 2019). Ulrich Spiesshofer, the former CEO of ABB, explicitly said that "to support our customers in a world of unprecedented technological change and digitalization, we must focus, simplify, and shape our business for leadership. Today's actions will create a new ABB, a leader focused in digital industries" (Vavra, 2019, p. 10).

Communication between human-to-machine and machine-to-human, as we know it from Industry 3.0, will change dramatically. With the advent of automation and digitization in the manufacturing industry, the need for machine-to-machine communication increases. One of the main features of IoT is that machines and devices exchange information without human intervention. The key challenge is integrating services effortlessly over the internet with active participants who exchange data about themselves and their nearby environments over a network-based infrastructure. The IoPTS, for example, is characterized by its vast sum of data compiled by services, things, and people that will constantly generate data, populating massive databases.

There is also an advanced potential for tracking people, things and objects that needs to be managed. This means that there are various aspects to consider; a person might have multiple identities where every identity is associated with numerous things (devices) and linked to different services and processes (Jeyanthi, 2018). All this might lead to significant challenges from a governance perspective and also demand for sound business strategies. Depending on who is talking to whom, it is necessary to respect languages, capabilities to read (get information), write (send information), save (storage information), have the competence to think, act autonomously, to answer, and many more functions. Still, it is essential to manage that amount of information and gain insights that enhance quality and internal processes, and benefit the clients. If not managed properly, companies might "drown in the flood of data." New approaches to business analytics need to be implemented (Silva et al., 2021) and adapted to the organizations' decision-making process.

Technical development might focus increasingly on strategies for Smart Homes, Smart Buildings, Smart Cars, Smart Factories and Smart Cities, whereby "smart" solutions refer to technical services which use sensors, controlled engines, or cameras with a focus on comfort, increased functionality,

FIGURE 1.1
Drowning in the flood of data. (It is time to learn to swim.)

power saving or security. The benefits as well as risks of smart services must be pondered between user experience and data security.

Heavy work might be shifted to robots through the focus on automation and digitalization of the manufacturing process of goods and services. Big data to be stored in huge computer centers allows devices to get information from anyone, everywhere, and to respond to requests for information in real-time. In addition, organizations have enhanced these industrial needs of transparency, efficiency, and intelligence of processes focusing on smart factory strategies. The first-level goal is to combine systems to control place, time, and orders for more transparency in real-time. This is followed by the second level strategy to connect machines indirectly over sensor technologies (Human, 2017).

- Machines, devices, and sensors network together
- Networking between machines, devices, sensors

For example, the International Federation of Robotics counted in 2017 a total of 389,000 traditional industrial robots, and about 11,000 collaborated robots. By 2020 the conventional industrial robots decreased to 362,000 units. However, the collaborating robots increased to a total of 22,000 units (Bieller, 2021). Apparently, organizations continue to transform their business models and production processes into data-driven strategies. The number of machines, devices, and sensors that collect structured and non-structured data from the real world will increase massively. With the changes in data-driven process, management in all production sites, branches, and organizations might collect data from anywhere and everyone. Sensors collect data from the real world and transmit that data to machines in real-time. This implies that the amount of generated and replicated data will increase yearly in the foreseeable future. The challenge is to make sense of this data to enhance automated production processes and extract relevant information for management and strategic decision-making.

Cloud Computing

Cloud computing is both a precondition and a side-effect of the new accelerated business environment. Cloud services can also be classified by the usage of the data which will be stored. In public clouds, every user can use the hardware, content, and services. It is possible to access all shared data within that cloud infrastructure. The new transformed and digital ecosystem will also bring complex security and privacy challenges. Fundamentally, there is a high demand for flexible, user-friendly ways to authenticate users. Without adequate management of digital identities, organizations will not only struggle with present problems, including identity theft, spam, malware, and cyber fraud, but they will also be unable to assure individual users that they can safely move essential data and applications from their local computers to the cloud. The opportunities offered by new digital possibilities might be missed (Cavoukian, 2008).

One of the new concepts of digital transformation is the idea of optimizing the efficient use of physical resources by sharing these resources. Huge companies and cloud infrastructure providers, like AWS, Azure or Google Cloud, give access to their machines through cloud services and open this infrastructure for a collaborated usage. Small and middle-sized companies can use the resources on a shared basis. The provision of infrastructure-as-a-service allows the use of shared virtual machines or infrastructure services. This allows vast scalability and additional resources whenever needed for the production or to satisfy clients' needs. This instant access to additional resources is sometimes called "rapid elasticity" or "infrastructure scaling" and requires adequate resource estimation from the provider or operator of a virtual data center (Bulla et al., 2020). A provider administers and maintains the business model platform-as-a-service infrastructure. Users have the ability to run their own freely designed programs on a platform and just pay for the time taken and resources used. With this approach, companies can reduce the cost of managing their IT resources and focus on developing new products and services for their customers. They might provide access to their products or services through a web application as a software-as-a-service (SaaS) offering. Clients can, in that way, more efficiently use the vast services even at a reasonable price.

Restraining Reasons, What About Data Safety?

Depending on the business model, it might be crucial to collaborate with partners and shared spaces for smooth teamwork in a collaborative environment. The adequate sharing of digital information with all parties involved in a supply chain can be considered a central prerequisite for successfully implementing Industry 4.0 strategies (Müller et al., 2020). Even if humans or machines work together, the information exchange needs to respect compliance and trust (Ryan & Falvey, 2012). It is essential that every partner needs to have the necessary information at hand for the current process step. Hence, organizations and data owners might need to invest in their information management and enhance their knowledge about data governance for streamlined processes without unnecessary risk exposure. Cachin and Schunter pointed out that it is necessary to protect one's own resources from the decryption of unauthorized others (2011). Many people fear the accessibility of data in the cloud and the loss of control. So far, we must keep in mind that we need to take care of all components, gateways, devices, and programs in our own infrastructure. It can be argued that, in particular, small and mid-size companies might have fewer resources to implement and maintain security mechanisms than specialized cloud providers.

The consulting company Fortune Business Insights points out that "data security and privacy concerns about data loss, data breaches, unforeseen emergencies, application vulnerabilities, and online cyberattacks allied to cloud-based solutions are expected to hinder growth" (Fortune Business Insights, 2021). This applies to a wide range of industries and sectors, and not without a good reason. According to a recent survey with more than 3,600 participants (IT and security professionals) around the globe, about 51% of the participants reported that their organizations had sustained a data breach in the last 12 months (IBM, 2021). Research has shown a high correlation between social interaction, trust, and benefit-sharing (Müller et al., 2020). Hence, an appropriate data governance mechanism for data access is crucial to sustain and thrive with new products and services in the transformed digital economy. Managing access to data is probably already now a "mission-critical" task for many organizations. A clear strategy and management focus can help to address the challenges in this context.

Collaborative Services with a Need for Speed

Globalization, collaboration over cloud services, massive computer centers, and a high frequent polling rate of data increase the need for speed. With cloud computing and new methods of Industry 4.0, we can learn to grow small ideas into valuable strategies for better collaboration. The amount of data we collect and transfer is enormous. To give an easy example of the evolution of required bandwidth, we can simply compare video sizes per minute and the need for resources for future solutions. A 720p HD video with 30fps requires a transfer rate of 60MB per minute—this was considered good quality video for a couple of years—now, a 1080p Full HD video with 60fps transfers already 175MB per minute, and with a 4K HD video with 60fps grows the transfer rate up to 400 MB per minute. For a single video of 10 minutes, we need to store up to 4GB. This rough calculation does not consider that there are probably also many more videos available. This means both the number of videos available as well as the size of a single video require more storage space. However, the 4th IR comes not only with video data but all kinds of data are produced with services like the IoT and the always-online culture.

Cloud-based services store and distribute a vast amount of data. This data is often replicated or copied, and all of this increases the demand for immense fast connections. We will consume tremendous bandwidth when discussing the IR and future solutions' growing cloud services. In Germany, for example, in 2020, the total internet transfer volume increased by approximately 100 billion GB, just from internet surfing (DPA, 2022). How would we measure the worldwide usage, and with what number would we have to multiply that demand in the future? Anyhow, the idea of video streaming is helpful for the understanding of timely related data we use. Focusing on future solutions like autonomously driven cars, we realize that a single car will collect data in the range of terabytes in an hour (Nelson, 2016). Millions of vehicles traveling over our streets daily increase interest in fast communication and data exchange. Each car has to collect data, make decisions in seconds and distribute and contribute data while computing real-time information simultaneously. An autonomous vehicle can be understood as a multi-sensor system.

Artificial Intelligence and Multi-Sensor Systems

Many methods, including those mentioned above, are also empowered by the using AI services. AI is a rather broad field. It includes subfields like deep learning, machine learning, and neural networks—which can be seen as the backbone of deep learning algorithms (Kavlakoglu, 2020). AI can support us from relatively simple services to very complex multi-sensor systems. Decision-making is usually based on a digital framework that combines algorithms and processing with fast connectivity services. "Weak AI" services, sometimes also called "narrow AI," are designed to perform a specific function (IBM, 2020); for example, self-driving a car, reacting to human speech, or textual requests. Weak AI relies on human input and needs to be supplied with relevant training data to advance learning and increase accuracy. Chatbots, for example, can support answers to customer requests on product information or support reasons. Virtual assistants and speech computers (like Alexa, Bixby, etc.) listen to whole communication in their proximity and provide interactive services in a Smart Home. This kind of system collects big data about users' preferences and enriches it with available information from the internet or cloud services. It can play music, answer simple questions, or change room conditions.

"Strong or super-strong AI" systems understand and compute from real-time collected big data sources. "While human input accelerates the growth phase of Strong AI, it is not required, and over time, it develops a human-like consciousness instead of simulating it, like Weak AI" (IBM, 2020). For example, the autonomous car needs to know about the internal situation of all sensors and the surroundings, like the outside environment and road conditions. Currently, most approaches in this context can probably be classified as "weak AI" as they depend on human interaction, training data, and testing. The sensors collect vast amounts of data about the measured environment. At the same time, the various car functions like speed and driving direction, need to be controlled and managed. However, the field is developing quickly, and autonomous cars might overcome that hurdle (dependence on human input and interaction) with more mature technology, and self-learning strategies will take over. Then it can be considered as an application based on "strong AI" (IBM, 2020).

The strategies for smart cars, buildings, cities, factories, and more need to be designed carefully. While the technology is often already quite mature, there is still work in progress to determine which new strategies are suitable for making the best use of this technology. We need to find suitable business models and understand the relevant supporting parameters (culture, guidelines, processes, knowledge, etc.) to make the best of it in a specific organization or business environment.

Cyber-Physical Systems Make Autonomous Work Possible

Cyber-physical systems (CPS) connect mobile and mechanical components (like machines, facilities, equipment, and robots) over networks with other systems. They are designed to process data in the whole production process. In that respect, IR combines both machine-to-machine communication and industrial internet of things for improved connectivity and operational efficiency (Kevan, 2019). Sensors collect data from the physical world and transfer it to an IT system for persistence and further processing. Big data exchange often happens over the internet and cloud computing, and the software analyzes the data. Finally, an algorithm computes possible decisions to operate physically embedded actuators; for example, track switches, locks, robots, windows, or doors (Luber, 2017). CPS can provide new ways of collecting and analyzing very fine granular levels of data throughout the production process. That data can also be used to enable predictive maintenance and big data analytics (Kevan, 2019).

Augmented Reality, Enriched Virtual Images of the Real World

Virtual reality concepts support the economy, medical sector, industry, and manufacturing services. Sensors, like cameras, can detect precise information in real-time and send the data to an augmented reality (AR) app. The AR app creates a virtual image of the object in a 3D space from this data. The (augmented) 3D view can be seen with smartphones or AR glasses. With the motion-tracking capabilities of mobile devices, the three-dimensional position in the (virtual) room can be determined, and the object can be viewed from different perspectives—even if the object in the real world is difficult to access. Some AR services are combined with AI logic and allow tracking or navigation on a real-time map (Human, 2019).

TECHNICAL EVOLUTION ENFORCES CHANGES IN BUSINESS ENVIRONMENTS

Challenges of Changing Working Strategies

Organizations cannot switch into the new world with a single mouse click. The transformation of organizations requires comprehensive change of efforts. Developing the metamorphosis from the traditional best-practice methods to the next business level will grow iteratively. Each business process needs to be checked in detail for new digital components and service changes. The business process management needs to take care of each process's design, infrastructure setup, and agential logic (Baiyere et al., 2020). With IR business models, we are bringing digital infrastructure and applications into clouds. We are connecting physical and virtual technical devices. That impacts a significant number of workflows, services, and business environments directly. In order to take care of the required hard skills (developing changes in infrastructure, process management methods, and services) and the soft skills (who will do what and where), clear responsibilities for the design and development of all relevant workflows need to be established. Even when many technical devices work together, we have to ensure that the operators and manager in charge of the process can understand what is going on and that they have complete control of the environment. Employees who have already worked digitally will probably adapt to these changes more quickly. For others, who have worked for a long time in non-digital workflows, it might be difficult to adapt to the job requirements. Through business changes, future collaborative workflows might be understood as a service.

To develop workflows further with the new strategies and methods, in particular, we need to take care of increasing communication needs. In its basic form, communication is just an information exchange from a sender to a receiver, and if both speak the same language, it might work.

From the view of human-to-human communication, we can mention that the quality and quantity of the information must not be perfect. Native speakers or experts can usually work without significant issues without perfectly presented information. Humans can estimate information and fill gaps with their own experience. In the communication with or between machines, it might not be that easy because devices need strict instructions and specific details on what it is necessary to do.

CONSEQUENCES FOR THE PEOPLE

We discussed the intense technological developments in this chapter, but we should not forget that it is also about people and how they perceive and interact with them. There are many aspects to this. From many perspectives, the world has grown into a smart world; this is great. Work seems to become more manageable. We can work from everywhere and anytime, which is turning out to be fruitful for many managers and global-working ideas. With that plus point, a few might also fight this idea because digitalization might kill or change the job they love. Automation is a commodity that disvalues this (small) group of people.

It also affects consuming information, learning management, and communication in teams or decision-making responsibilities. The IoP is a concept that summarizes some of the ideas in this context. It brings human behavior and algorithms together. It is about learning and human experience. It can also be considered a paradigm shift from platform-centric to human-centric approaches (Conti & Passarella, 2018). Social networks facilitate interactions and monitoring of social relationships, users' social degree, friendships, recommendations, etc. (Mumin et al., 2022). The IoP might not be a different technology, but it is a way we use the new services on a global scale. We might need to adopt new strategies to deal with the changes and leverage the technology in our day-to-day business. Andrea Lipton suggests five key elements to capitalize on the IoP (2017). She states that, in order to use the full potential to embrace learning and development, you should be:

- Social. Focus on connecting people instead of content.
- Personal. Focus on individuals, not the job role.
- Proactive. Do not wait. Go! Build relationships before you need them.
- Predictable. Be specific about success and expected behaviors. Unleash employees.
- Porous. Stay open to ideas to see what works.

Lipton also emphasizes that "the ever-increasing pace of innovation and constant emergence of new technology has made competencies like risk-taking and tolerance for ambiguity critical" (Lipton, 2017, p. 49). This changed paradigm might imply a new mindset and most probably requires adaptations to the organizational culture.

FIGURE 1.2
Changed organizational culture: established and new methods need to be harmonized.

But the development does not stop at the IoP. We can think about further integration and connectivity. The IoPTS, for example, can be seen as a metaphor where people, things, and services seamlessly interact over the internet. People and machines (algorithms) can be seen as active participants in data exchange. This data might be about the current state but also the nearby environment. Some people might even consider it a flood of data, and we need clever strategies to make the best use of it. The IoPTS comes with many features; however, it is also characterized by its enormous sum in terms of services, things, and people that will generate data. It also comes with an advanced potential to track people, things, and objects (Jeyanthi, 2018). Overall, the side-effects of the IoPTS might significantly increase complexity. Besides, with already mentioned changes in mindset and organizational culture, there might also be changes in governance necessary, not to forget the importance of the strategic perspective.

CONSEQUENCES FOR OUR COMMUNICATION

The always-on culture impacts the communication of society as a whole, including business communication and private interactions in many situations. In this respect, employees' expectations might be identical to those of clients. The experience must be tailored to specific individual requirements and preferences. That means the tasks, the coaching, and the manager in charge should have an open ear for each employee's unique requirements (Lipton, 2017). While it is disruptive to various established communication patterns, it can also be considered an enabler for social contact. The "online world becomes more familiar and interesting" (Lüders & Gjevjon, 2017). Ultimately, we might need to get used to the new communication patterns and establish routines to deal with the increased complexity.

SUMMARY

The opportunities that come with the vast implication of how Industry 4.0 will influence our communication and work environment can probably not be overestimated. Some authors see Industry 4.0 as the combination of the IoT and the IoP (Jeyanthi, 2018) but it is probably much more than that. The IR also includes a significant cultural paradigm shift. Managers and other decision-makers need to be aware that the technology is here and will impact the work environment in many industries, whether you like it or not. It certainly brings lots of opportunities and chances for new business. But it will not come without side-effects that have to be managed. The organization and the people acting in the organization have to be prepared and trained. Appropriate processes, policies, and governance must be established within the organization and society as a whole. This challenge should be among the top priorities for decision-makers and strategy boards, but it is also a question of organizational culture and the mindset of the workforce to ensure people do not "drown in the flood of information," do not get overwhelmed with the "always-on-culture", and to make the best use of the rich opportunities given with the latest technology available.

REFERENCES

Baiyere, A., Salmela, H., & Tapanainen, T. (2020). Digital transformation and the new logics of business process management. *European Journal of Information Systems, 29*(3), 238–259. https://doi.org/10.1080/0960085X.2020.1718007

Bendel, P. D. O. (n.d.). *Definition: Industrie 4.0* [Text]. https://wirtschaftslexikon.gabler.de/def inition/industrie-40-54032; Springer Fachmedien Wiesbaden GmbH. Retrieved April 27, 2022, from https://wirtschaftslexikon.gabler.de/definition/industrie-40-54032

Bieller, D. S. (2021). *Director Statistical Dpt. Vice Chair IFR Service Robot Group. 43.*

Bulla, S., Reddy, C. V. R., Padmavathi, P., & Padmasri, T. (2020). Analytical evaluation of resource estimation in web application services. *Ingénierie Des Systèmes d Information, 25*(5), 683–690. https://doi.org/10.18280/isi.250516

Cachin, C., & Schunter, M. (2011). A cloud you can trust. *IEEE Spectrum, 48*(12), 28–51. https://doi.org/10.1109/MSPEC.2011.6085778

Cavoukian, A. (2008). Privacy in the clouds. *Identity in the Information Society, 1*(1), 89–108.

Chhaya, K. (2020). Convergence of 5G, AI and IoT holds the promise of Industry 4.0. *Telecom Business Review, 13*(1), 60–64.

Conti, M., & Passarella, A. (2018). The Internet of People: A human and data-centric paradigm for the Next Generation Internet. *Computer Communications, 131*, 51–65. https://doi.org/10.1016/j.comcom.2018.07.034

DPA. (2022, March 28). Viele, viele Gigabytes: Internet-Datenvolumen wächst weiter rasant an. *Die Zeit.* www.zeit.de/news/2022-03/28/internet-datenvolumen-waechst-weiter-ras ant-an?utm_referrer=https%3A%2F%2Fwww.google.com%2F

Fortune Business Insights. (2021, October). *Cloud Computing Market Size, Share & Industry Growth [2028].* www.fortunebusinessinsights.com/cloud-computing-market-102697

Gupta, P., & Goyal, K. (2021). Change management in Industry 4.0-based organizations. *Change Management: An International Journal, 21*(2), 47–63. https://doi.org/10.18848/2327-798X/CGP/v21i02/47-63

Healy, W. (2018, November). What's the difference between Industrial IoT and Industry 4.0. *Hydraulics & Pneumatics.*

Human, S. (2017, December 5). *Vernetzungsstrategien für die Smart Factory im Vergleich.* www.industry-of-things.de/vernetzungsstrategien-fuer-die-smart-factory-im-vergleich-a-1077474/

Human, S. (2019, November 20). *Augmented Reality in der Industrie: Herausforderungen, Potenziale, Chancen.* www.industry-of-things.de/augmented-reality-in-der-industrie-herausforderungen-potenziale-chancen-a-882695/

IBM. (2020, August 31). *What is Strong AI?* www.ibm.com/cloud/learn/strong-ai

IBM. (2021, November 16). *Cyber Resilient Organization Study 2021.* IBM. www.ibm.com/resources/guides/cyber-resilient-organization-study/

Jeyanthi, P. M. (2018). INDUSTRY 4.O: The combination of the Internet of Things (IoT) and the Internet of People (IoP). *Journal of Contemporary Research in Management, 13*(4), 29–39. Business Source Complete.

Kavlakoglu, E. (2020, May 27). *AI vs. Machine Learning vs. Deep Learning vs. Neural Networks: What's the Difference?* www.ibm.com/cloud/blog/ai-vs-machine-learning-vs-deep-learning-vs-neural-networks

Kevan, T. (2019, November 1). Making sense of industrial connectivity trends. *Digital Engineering.* www.digitalengineering247.com/article/making-sense-of-industr ial-connectivity-trends

Lipton, A. (2017). The Internet of People delivers new ways of learning. *People & Strategy,* *40*(3), 48–51.

Luber, S. (2017, December 5). *Was ist ein Cyber-physisches System (CPS)?* www.bigdata-insider. de/was-ist-ein-cyber-physisches-system-cps-a-668494/

Lüders, M., & Gjevjon, E. R. (2017). Being old in an always-on culture: Older people's perceptions and experiences of online communication. *The Information Society, 33*(2), 64–75. https://doi.org/10.1080/01972243.2016.1271070

Müller, J. M., Veile, J. W., & Voigt, K.-I. (2020). Prerequisites and incentives for digital information sharing in Industry 4.0 – An international comparison across data types. *Computers & Industrial Engineering, 148,* 106733. https://doi.org/10.1016/j.cie.2020.106733

Mumin, D., Shi, L.-L., Liu, L., & Panneerselvam, J. (2022). Data-driven diffusion recommendation in online social networks for the Internet of People. *IEEE Transactions on Systems, Man, and Cybernetics: Systems, 52*(1), 166–178. https://doi.org/10.1109/TSMC.2020.3015355

Nelson, P. (2016, December 7). *One autonomous car will use 4,000 GB of data per day | Network World.* www.networkworld.com/article/3147892/one-autonomous-car-will-use-4000-gb-of-dataday.html

Ryan, P., & Falvey, S. (2012). Trust in the clouds. *Computer Law & Security Review, 28*(5), 513–521.

Schwab, K. (2016, January 14). *The Fourth Industrial Revolution: What it means and how to respond.* World Economic Forum. www.weforum.org/agenda/2016/01/the-fourth-industrial-revolution-what-it-means-and-how-to-respond/

Silva, A. J., Cortez, P., Pereira, C., & Pilastri, A. (2021). Business analytics in Industry 4.0: A systematic review. *Expert Systems, 38*(7), 1–26.

Vavra, B. (2019, January). Industrial Internet groups to collaborate on cloud computing, IIoT research. *Control Engineering,* 10.

2

How Do You Keep Your Employees Engaged in Remote Work?

Stephan Meyer
Almademey Ltd, Cyprus

CONTENTS

DOI: 10.1201/9781003371397-2

DO YOU LIKE THE CURRENT SITUATION?

The Traditional Office Is Dead. How Do I Keep Remote Employees Engaged?

We thought we had more time to adjust to the inevitable transition from traditional working arrangements from 9 to 5. As leaders, we read statistics which claim that the economy is about to change in the near future, but we read this information with some degree of rejection. Over the time, business was going well, and every effort was made to keep employees motivated and productive. Now we have to get used to the new standard: Remote work is working from anywhere in the world, assuming you have access to electricity,

FIGURE 2.1
Remote work.

internet, and other necessary resources. Remote work, or work from home (WFH), as some people call it, is here to stay. Even conservative estimates say that 20% of full workdays will be supplied from home after the pandemic, compared with just 5% before (Barrero et al., 2021).

A PANDEMIC THAT TRANSFORMED THE WORKING WORLD OVERNIGHT

COVID-19 came, much to our dismay, and we were forced to move our businesses quickly just to survive. We have moved hundreds of employees from traditional office jobs to work remotely. Some were unable to do so and lost their jobs. Others who were considered indispensable workers had to abide by a whole new set of frightening rules. Not only has it overwhelmed employees but it has also consumed every ounce of our energy to meet daily demands and recommendations of everyone, from the medical community to politicians. It comes as no surprise: For some people the pandemic was a catalyst for self-reflection and a questioning of their own lifestyle. "Covid has not been all doom and gloom for me. It forced me to dig deep and reevaluate what is important" (Microsoft, 2022).

Pre-pandemic research from 2019 sought to determine why there are still people who do not work remotely (Lott & Abendroth, 2020). Significantly more men than women viewed their jobs as unsuitable for remote work because of the manual nature of the work. Admittedly, this is a good reason. However, significantly more women than men, albeit at a much lower percentage, viewed their work as suitable for remote work but were still expected to not work remotely due to perceived cultural barriers. "Cultural barriers" is a euphemism for superstition.

Fast forward to today. Our organizations have changed forever. It may come as a surprise to some. Still, more often than not, participants in studies (Isac et al., 2021) are satisfied with the current trend of remote work altogether, and they have also gravitated toward continuing WFH even after the pandemic period ends. Some employees equate remote work with a pay raise (Barrero et al., 2021). In other words, it is no longer about planning remote work but about making adjustments to this new way of working. One of the biggest challenges is maintaining a regular connection with employees. Creating a culture where everyone works in the same area is easy but, when everyone works from their own location, that culture often becomes a problem. With uncertainty about when or if we will ever return to the traditional office environment or whether remote work has replaced it forever, how can we foster this important sense of culture and belonging in our employees?

Remote Work Is Becoming the New Norm

The pandemic is revolutionizing the workplace. The least that can be said about it is that it may have helped put things in a different perspective. As one employee put it: "Work is only a part of life. It shouldn't be your whole life or the only thing you care about" (Microsoft, 2022). Teams are fundamentally changing the way they communicate and work. It is difficult to estimate the number of companies that have worked remotely since the start of the crisis, but Gartner experts provide some clues in a survey from March 2020 (Gartner, 2020), at the beginning of the Covid pandemic. The team interviewed 800 human resource (HR) leaders from around the world and found that 88% of companies encouraged or required employees to WFH, regardless of whether they showed coronavirus-related symptoms or not.

Remote work is dramatically changing how employees interact with the rest of the team and their managers. To help employees maintain productivity levels and not feel disconnected, companies are creating additional virtual records for employees with managers and introducing new tools for virtual meetings. In fact, running the business as smoothly as possible is a priority for most, if not all, managers. But are these measures sufficient to avoid interruptions in activities during a crisis? The thing is that employees WFH full time until further notice. In addition, this new way of working has been implemented unchanged, making remote work a major challenge for both companies and employees.

HOW DO WE MAINTAIN ENGAGEMENT WHEN OUR ORGANIZATIONS HAVE ALREADY BEEN STRUGGLING?

Let us look at what corporate culture and commitment consist of. These ideals focus on the values of people moving toward a common goal. What could have started little by little became a kind of movement. More people got on board and believed in it, so a culture was born. Employees can choose whether they want to stay connected or exist on the boundary in a non-activated state. And, statistically, about two-thirds of the workforce did so.

The same employees will likely remain employed in your organization, along with several new faces. It is important to take advantage of those on board and take into account the values you have established over time. It is likely that these marginal employees are already doing their thing or have switched to other companies. Keeping employees engaged will be a major challenge for tomorrow's leaders. The future of leadership has a new label: "Leading from home" (Antonacopoulou & Georgiadou, 2021).

HOW CAN YOU RESTORE A SENSE OF ENGAGEMENT IN REMOTE EMPLOYEES?

The secret of employee engagement has always been a mystery. There is no hard rule that works for all organizations. Instead, each organization must take the time to learn what drives its employees and boosts engagement. Much of this comes down to improving the workplace so that people feel valued and respected and that their talent contributes to greater well-being.

FIVE PROPOSITIONS ABOUT WORKING REMOTELY

Proposition 1: Working Remotely Is Inherently a Good Thing

What are the benefits of working remotely? Beginning with the major challenge of most HR departments, the acquisition of qualified employees: Companies can hire and retain the best talent from anywhere while avoiding the inconvenience that comes with relocation and cultural adjustment. Also, companies reduce facility costs. Renting an office can be a major cost factor for any organization. If you go so far as to buy an office building, it means you are entering the real estate business. In contrast, remote work means that employees get to work anywhere at any time and can collaborate more effectively with their team members. Employees usually appreciate being able to avoid the hassle and expense of commuting. Some remote workers take full advantage of the opportunities that a remote working lifestyle gives them. In order to open their minds, achieve greater global understanding, and expand their professional network into a worldwide community, they take their remote work to different countries around the world, either through work and travel programs or DIY travel arrangements.

As a rule, working remotely has many benefits, including the potential for increased flexibility and productivity. Working remotely can be a great way to improve work and life balance. Remote workers are less likely to feel overwhelmed or stressed out at work. By working remotely, you can avoid possible conflicts with coworkers and clients, which can lead to better communication and overall teamwork. Admittedly, remote work also poses a few new challenges from IT support to internet connectivity to data privacy. Nevertheless, the positive aspects are likely to outweigh the negative if a prerequisite described in the subsequent paragraph is met.

The prerequisite for remote work being more productive is that remote workers know how to organize themselves in a way that leads to fewer interruptions as compared to the ones experienced in a classical office environment. A study (Iqbal & Horvitz, 2007) found that participants spent, on average, nearly 10 minutes on switches caused by alerts, and spent, on average, another 10 to 15 minutes (depending on the type of interruption) before returning to focused activity on the disrupted task. Sometimes the interruption would animate the users to temporarily focus on other tasks so that 27% of task suspensions resulted in more than two hours until resumption. Therefore, a smart and more focused way of organizing one's work may provide a productivity benefit of remote work over office work. A part of this is a dedicated working space for remote work. A Japanese study (Takahashi, 2021) put it bluntly: "Unless there is a private room with office facilities (similar to an office environment) at home, household and work responsibilities interfere with each other, and thus, one cannot work efficiently."

Proposition 2: Productivity Moves to the Cloud

Cloud-based technologies enable communication and interactions between workers without the need for direct physical presence. As a plus it is much easier to form teams across locations based on specialist tasks—without killing the team leader with travel time. Enterprises are using apps and tools to improve the effectiveness of remote work, such as Slack for communication, Google Docs for collaboration, and Zoom or MS Teams for video meetings. You can say without exaggeration that every day new tools enter the market that are supposed to enable users to collaborate and communicate remotely. Employees are social beings—social interaction, meetings, concentrated work; there are solutions in the cloud for all these problems—but they have to be introduced and adapted if necessary. In other words, remote work and cloud benefit each other but are not a foregone conclusion. However, what does it mean for the company culture? The increased responsibility for work may lead to an increase in hierarchical power. In any way, remote work is seen as a privilege in some companies or may even be considered a promotion.

Productivity moves to the cloud in order to improve collaboration and communication. The cloud allows employees to work from any location and eliminates hassles such as printing, copying, and faxing (the older ones will remember). This makes it easier for companies to attract talent by providing better job security and making remote work a more viable option. Employees can access their work files at any time, which improves efficiency and productivity. The cloud also reduces costs by eliminating the need for expensive office space and equipment. There are many benefits of using the cloud for productivity, including increased communication between employees and improved efficiency in terms of workflow.

All in all, the value generation process of an organization becomes less physical and more virtual, with only a few exceptions depending on the industry. The cloud enables a better division of labor through workflows— and, in the near future, one can then automate many decisions from the data collected in the workflows in the cloud. Once a modus operandi of working remotely has been found, even the most complicated tasks are suddenly possible. In the midst of a global pandemic, a Canadian public sector organization, for example, embarked on a reorganization, a new corporate strategy, orientation of a new executive team and a shift to a purpose-driven organization, all while 80% of its workforce was WFH or in the field (Gilpin-Jackson & Axelrod, 2021).

Proposition 3: Men and Women Experience Remote Work Differently

There seems to be a gender divide concerning remote work. According to a survey (FlexJobs, 2021) from March and April 2021, out of 2,100 people, approximately 68% of women said their preferred workplace, even after the pandemic, would be remote only, compared to 57% of men. An impressive 80% of women said remote-work options are among the most important factors to consider when evaluating a new job, compared to 69% of men. Both genders mentioned "no commute" as the leading benefit of remote work; 87% of the women and 74% of the men. An especially large discrepancy between the genders was that 70% of women claimed not having to get dressed up for work in formal office clothes was a benefit, compared with 57% of men. The Dutch Longitudinal Internet Studies for the Social Sciences panel came to results indicating that workers with a long commuting duration who transitioned to WFH increased their subjective well-being (Delft University of Technology, 2022). However, this effect was found to be significant only for women and not for men. In general, several sources indicate that remote work is more popular among women than among men (Beňo, 2021; Delft University of Technology, 2022; FlexJobs, 2021; Isac et al., 2021; Microsoft, 2022; Tudy, 2021).

Proposition 4: Companies May Lose a Competent Workforce

From the perspective of a business economist, a company is a bundle of contracts. Remote work transforms all employees into what feels like solo self-employed. This can lead to some professional benefits such as increased productivity, reduced costs, and unparalleled creativity. But it also carries some risks that should be considered so they can be mitigated. Most of all, this can be dangerous for the company.

When employees realize they can work alone, they may like it so much that they would want to become self-employed. Independent contractors are individuals who provide services to clients, usually in the form of contracting with a business. For small companies, this is risky. In this case, a freelancer can offer high-quality services at a lower cost and their own set of rules. Freelancers have more options than other employers due to flexibility and freedom. Even if they have fewer company benefits, this can still be a very tempting lifestyle model for some employees. So, a loss of competent human resource with the consequence of costly hiring procedures for their successors could be a possible result of this effect. For large companies, however, losing competent employees to freelancing is less of a risk because these companies are capital intensive. It is much harder to replicate a large company as a solo entrepreneur. Smaller and larger companies are affected differently by the trend of remote work. According to a study from India (Silva, 2021), the transition to remote work proved more challenging for small and mid-sized companies compared to large companies. For example, a larger proportion of small companies struggled to secure remote work authorization from clients compared to mid-sized and large companies.

Proposition 5: Remote Work Is Also a Risk for the State

Those who can take good care of themselves economically are less susceptible to socialist promises of salvation. Rather than becoming dependent and believing that the government and its inherent socialism will take care of their needs, the most competent members of the workforce will resort to geo-arbitrage. If you are dealing with the topic of geo-arbitrage for the first time, then you have certainly stumbled over the peculiar term at first. Arbitrage is French and comes from the Latin word *arbitratus*. This in turn means something like "free choice" or "free discretion." In other words, it is about someone moving to a place where the general conditions of living are much more favorable than in their former home country. This could be due to factors such as less bureaucracy, a lower cost of living, better weather, a more relaxed local lifestyle and *l'art de vivre*, and last but not least, lower taxes. If you come from a country such as Germany with the highest taxes in the world, any relocation to another country is an improvement. A country that does not offer good general conditions for the productive part of its people will suffer from a massive brain drain, losing taxpayers and thus their most important source of income. The author of this chapter would argue that some countries have made a concerted effort to close this gap by setting up attractive governmental programs for entrepreneurs. Other national governments, such as Germany's or Belgium's, have not even begun to comprehend and address this problem.

SO, HOW CAN YOU IMPROVE REMOTE WORK?

Here is a list of hints and suggestions to make your organization more productive when working remotely. This list does not claim to be complete.

Provide Opportunities for Informal Communication

In physical life there are several formats that enable informal communication beyond the usual formal communication (such as in meetings). These formats include hallway conversations, joint visits to a canteen or restaurant, company barbecues, and perhaps even a company outing. Since informal communication plays an important role in exchanging ideas across hierarchies and bonding among employees, opportunities for informal communication should also be found in the virtual world of remote working. Imagination is needed here to develop formats that fit the company. For example, the author of this chapter knows of a company where the boss lets his employees play geocaching competitions. The employees are really looking forward to seeing their colleagues again in this way, outside of pure workplace collaboration. There is also plenty of scope here for the moderators of online events to create opportunities for informal communication. This will certainly play a greater role in the training of moderators in the future. Some virtual conferencing apps have already pioneered this approach, creating spaces in the business context for chance encounters and small talk with previously unknown people.

Make it Easy for Employees to Focus on the Work They Really Love

No one joins a company because they love emptying their inbox in the morning. Administrative work is a necessary part of any job, no matter how specialized it is. However, employees bogged down in repetitive tasks might start to forget what they love about their profession. It is estimated that the average office worker receives around 121 emails every workday, including spam emails (PSP IT Design & Development, 2021; Templafy, 2020). A much lower number, but still disturbingly high, can be found at Atlassian. According to this source, employees receive, on average, 304 business emails per week (Atlassian, n.d.). If you now consider the 10–15 minutes it takes to refocus (Iqbal & Horvitz, 2007) after handling incoming mail, this is a cry for help for a more efficient way of office communication that does not just rely on email.

Make Sure Everyone Knows They Are Part of the Team

One of the hardest parts of managing remote workers is simply remembering to involve them. It is just as important to engage remote workers in team meetings and acknowledge their contributions publicly as it is for local employees. Resist the temptation to prevent remote workers from scheduling ad hoc meetings quickly because it is too tedious to let them know; you should have a quick and easy means of communication. Equally important when assigning tasks, brainstorming, or discussing solutions to problems, remember to involve remote workers in the process. Otherwise they will be isolated and cannot contribute effectively to the company's performance. Including everyone applies specifically to those employees who are new to the team. They will need the most attention from you.

Give Your Employees More Responsibility, Not Just More Tasks to Do

Working in a company that values morale can give you some exciting ideas on how you can engage employees. It is especially easy to give employees real responsibility at a time when they are incredibly dedicated to their work when "flow" conditions seem to be abundant. This should not be confused with giving your team more things or tasks. Instead, this gives them important projects and initiatives to take responsibility for and get out of the way. People are naturally goal-oriented. So, if you give them something worth achieving, you will be amazed at how much sense and drive it gives them. If you are a manager, give one of your direct reports an important project in which they can take the lead. Preferably, ask your employee to undertake a project that you think will help them grow and learn something new. Even if they have no experience in it, if they feel that this goal is important to the success of the company, they will find a way to achieve it.

Communicate Company Goals

Remote teams need regular communication to help them align with company goals. When you communicate changes to the overall strategy, you give remote employees the opportunity to adopt the new direction. This helps them become more productive and add real value to your business. If your strategy is in the process of development, the order in which you make changes to your overall strategy is important. The sooner you can implement the changes, the more employees will see them as new and exciting. The later in the process, the more employees will be skeptical about change. The better managers can communicate a sense of purpose to their team, the better the results will be (Ariely et al., 2008).

Set Expectations

Managers should work with team members to set realistic expectations for remote work. Team size, individual capacity, deliverables, schedule, and overall performance goals are discussed. The agreement is reached on how to communicate effectively. The more your employees know what is expected of them, the more efficient and productive they will be. On-site managers should be prepared to reach out to remote team members in times of need.

Issues like performance and budget adjustments require quick action. Regular check-ins allow remote employees to know what is expected of them and coach each other as needed. A healthy work environment is a result of regular communication across time zones. Team leaders cannot be left alone here—they need to be trained on how to lead virtual teams.

Create a Culture of Accountability

Create a culture of accountability in your organization by ensuring that your managers are held accountable for productivity and client satisfaction. Create a profit and loss culture by making sure that you are competitive with your competition and always stay profitable so that you can reinvest into new projects. Make yourself acquainted with the concept of "skin in the game": It is one that is increasingly being heard more and more as these turbulent times bring forth new opportunities and challenges. Be honest and ethical. If you cannot be honest or moral with your business practices, then how can you expect others to trust you? You need to build a reputation that follows you wherever and whenever possible.

Make Meetings More Productive

Ensure that meetings are more than just nice opportunities to talk to each other. Every meeting must have a result, ideally a list of to-dos with clear attribution about the task, the person responsible, and the expected deadline. Pro tip: If you are in a meeting and it starts to get off track, throw out action items until the meeting has a focus again. If people are not listening, it could be because they are not sure what you are trying to say. Implement a meeting timer. It will help to stay on track and ensure that everyone is prepared for their speaking time. One of the author's clients has a "camera on" company policy. The consequence: Participants stay attentive—people also do not interrupt each other so often. Other possible rules of the game: The inviter immediately reminds unpunctual participants that the meeting has started. Or the meeting duration is generally only 45 minutes so that you have some time between meetings.

Protect Your Employees from Digital Burnout

There is a prominent claim in the Microsoft 2022 Work Trend Index: Flexible work does not have to mean "always on." (Microsoft, 2022). The author of this chapter would subscribe to that. There are times when you expect an immediate response from your employees, and rightfully so, but there are also times when you do not. Your responsibility as a leader is to give your employees time to breathe. Too many online meetings, especially back-to-back meetings, quickly leads to waning energy and focus. An employee who sits in meetings all day can hardly be productive. Everybody needs some time in between meetings to refocus and to really produce results. A hot topic is also how to handle emails (or chats or other forms of communication) delivered outside working hours. Make it clear that you do not expect an immediate reaction. You may even use the label NOT URGENT in the headline of emails or chats when your colleague is in a meeting or it is outside working hours. Remember that VW switches off its mail server in the evening, while Daimler deletes correspondence that arrives during vacation (Kaufmann, 2014). You may not want to go this far, but you must acknowledge that this is how some companies handle this challenge. Things do not get any less complicated when you also take into consideration that more and more projects are delivered by employees working together in different time zones. There will be no solution that is a universally applicable solution. However, you are the one responsible for finding a solution that best fits your organization.

ONE FINAL WORD

So what is the gist of all this? Perhaps it is this simple realization: For the most part, there is no reason that everyone has to be in the same place when working together. Your focus should be on the customer, not the work. The customers are the ones who will help your product or service succeed. The way to get there is through satisfied employees. With a pleasant environment for remote work, you can achieve this goal more easily and elegantly.

REFERENCES

Antonacopoulou, E. P., & Georgiadou, A. (2021). Leading through social distancing: The future of work, corporations and leadership from home. *Gender, Work and Organization*, 28(2), 749–767. https://doi.org/10.1111/gwao.12533

Ariely, D., Kamenica, E., & Prelec, D. (2008). Man's search for meaning: The case of Legos. *Journal of Economic Behavior and Organization*, 67(3–4). https://doi.org/10.1016/j.jebo.2008.01.004

Atlassian. (n.d.). *You waste a lot of time at work.* Retrieved April 11, 2022, from www.atlassian.com/time-wasting-at-work-infographic

Barrero, J. M., Bloom, N., & Davis, S. J. (2021). Why working from home will stick. In *National Bureau of Economic Research.*

Beňo, M. (2021). From face-to-face to face-to-display management. *Advances in Business-Related Scientific Research Journal, 12*(1).

Delft University of Technology. (2022). Working from home during the corona-crisis is associated with higher subjective well-being for women with long (pre-corona) commutes. *Transportation Research Part A: Policy & Practice, 156*(Feb 2022), 14–23.

FlexJobs. (2021). *Survey: Men & Women Experience Remote Work Differently. FlexJobs.* Retrieved April 11, 2022, from www.flexjobs.com/blog/post/men-women-experience-remote-work-survey/

Gartner. (2020). *Encourage employees to work from home due to COVID.* Retrieved April 11, 2022, from www.gartner.com/en/newsroom/press-releases/2020-03-19-gartner-hr-survey-reveals-88--of-organizations-have-e

Gilpin-Jackson, Y., & Axelrod, R. H. (2021). Collaborative change engagement in a pandemic era & toward disruptive organization development practice. *Organization Development Review, 53*(2), 9–18. Retrieved from http://search.ebscohost.com/login.aspx?direct=true&db=a9h&AN=149985197&amp%0Alang=es&site=ehost-live

Iqbal, S. T., & Horvitz, E. (2007). *Disruption and recovery of computing tasks.* 677–686. https://doi.org/10.1145/1240624.1240730

Isac, N., Dobrin, C., Celik, B. K., & Azar, M. H. (2021). Does working from home influence motivational level of employees? The analysis of gender differences in Turkey. *458 Review of International Comparative Management, 22*(4), 459.

Kaufmann, M. (2014). *Erreichbar nach Dienstschluss: Maßnahmen der Konzerne – DER SPIEGEL.* Retrieved April 16, 2022, from Der Spiegel website: www.spiegel.de/karriere/erreichbar-nach-dienstschluss-massnahmen-der-konzerne-a-954029.html

Lott, Y., & Abendroth, A. K. (2020). The non-use of telework in an ideal worker culture: Why women perceive more cultural barriers. *Community, Work and Family, 23*(5). https://doi.org/10.1080/13668803.2020.1817726

Microsoft. (2022). Great expectations: Making hybrid work. In *2022 Work Trend Index: Annual Report.* Retrieved from www.microsoft.com/en-us/worklab/work-trend-index/great-expectations-making-hybrid-work-work

PSP IT Design & Development. (2021). *How many emails does the average office worker receive?* Retrieved April 11, 2022, from www.outlooktracker.com/news/how-many-emails-does-the-average-office-worker-receive/

Silva, R. J. R. (2021). Covid-19 pandemic and infrastructure management in technopark. *International Journal of Research in Commerce & Management, 12*(05), 5–9.

Takahashi, N. (2021). Telework and multi-office: Lessons learned from the bubble economy. *Annals of Business Administrative Science, 20*, 107–119. Retrieved from https://doi.org/10.7880/abas.0210705

Templafy. (2020). *How many emails are sent every day? Top email statistics for businesses.* Retrieved April 11, 2022, from www.templafy.com/blog/how-many-emails-are-sent-every-day-top-email-statistics-your-business-needs-to-know/

Tudy, R. A. (2021). From the corporate world to freelancing: the phenomenon of working from home in the Philippines. *Community, Work and Family, 24*(1), 77–92. https://doi.org/10.1080/13668803.2020.1809994

3

Remote Work and the Value of Informal Networks

Tobias Endress
Asian Institute of Technology (AIT) | School of Management,
Bangkok, Thailand

CONTENTS

"NEW WORK" IS NOT NEW… WHY IS IT STILL SO DIFFICULT?

The concept of New Work is not really as new anymore as the name might suggest. In fact, the term is already several decades old and was coined by the German philosopher Frithjof Bergmann. He can be considered the founder of the New Work movement. He used the term New Work to describe "the effort to redirect the use of technology so that it isn't used simply to speed up the work and in the process, ruin the world – turning rivers into sewers and rain into acid" (Bergmann, 1994). He also emphasized that "the purpose of technology should be to reduce the oppressive, spirit-breaking, dementing power of work – to use machines to do the work that is boring and repetitive. Then human beings can do the creative, imaginative, uplifting work" (Bergmann, 1994). In other words, he believed that "New Work is simply the attempt to allow people, for at least some of their time, to do something they passionately want to do, something they deeply believe in" (Bergmann, 1994). While this is a rather broad definition and it makes sense to have this in mind, the current focus is somewhat narrower. When it comes to the implementation of "New Work" concepts, most companies probably think about mobile technologies and remote work first.

Already in 2017, a survey from the consulting company Kienbaum revealed that mobile technologies (61% of the companies) and home offices (70% of the companies) are at the top of the list of measures companies are taking to introduce new work (Kienbaum, 2017). Many leaders say their organization plans to return to the office full time within the next year, but most employees prefer remote and hybrid work flexibility. Currently, the trend toward remote and hybrid work continues (Microsoft, 2022). It is a fair assumption to presume that this trend has increased in the Covid crisis. There is also strong evidence from research which indicates that remote work will stay (e.g., Leonardi, 2021; Sarkar & Kedas, 2022). Not only do employees want to stay at home, there are several indications that it is more productive (Awada et al., 2021; Rodrigues et al., 2022). At the same time, other authors suggest that productivity is higher in settings with physical presence in the office (Morikawa, 2022).

Overall productivity, efficiency and well-being in WFH settings probably depend on several factors, including skills and (growth) mindset (Rodrigues et al., 2022). That is why this chapter focuses on some of the aspects related to remote work, virtual teamwork, and WFH. While there is probably no way back in the old structure for most employees, it appears that the new structures are still work in progress in many organizations. In a recent study from the Microsoft WorkLab, for example, it was found that many hybrid employees (51%) are considering a switch in the near future to remote, while even more remote employees (57%) are considering a switch to hybrid work (Microsoft, 2022). Many employees are still adapting to the new normal and question existing processes and structures. Of course, this is not always an easy transition. There are some well-documented downsides of remote work, including "lack of environment change, lack of balance between work and personal life, lack of face-to-face communication with other employees, lack of inspiring working atmosphere and difficulty to stop working in the evening" (Simenenko & Lentjushenkova, 2021). Many business leaders emphasize that "relationship-building activities are the greatest challenge of having employees work hybrid or remote" (Microsoft, 2022, p. 43).

TYPICAL CHALLENGES OF REMOTE WORK AND HOME OFFICES

There are some excellent arguments for WFH, and it for sure has many benefits. This includes the possibility of reducing office space costs and allowing more expansive geography for recruiting employees. For employees, the benefits may include a better balance between work and personal life, reducing the need to commute, less emotional stress, additional free time, as well as the option to take on work in another region without the need to move (Blumberga & Pylinskaya, 2019). However, there are also some significant downsides that need to be managed. A recent study highlighted three main problem areas: Home office constraints, work uncertainties, and inadequate tools (Ipsen et al., 2021). Of those three disadvantages, the home office constraints are the most ambiguous as they consist of many different factors, but many of them are related to social interaction with colleagues. "We must not neglect organizations' role in supporting virtual, informal employee learning during the COVID-19 pandemic and beyond as remote work becomes the new normal" (Zajac et al., 2022, p. 283).

The networking within the company is of no value in itself, but there are various positive effects associated with it. Informal professional networks are positively related to increased discretion and proactive organizational culture. In the absence of other options, professionals use their informal networks to be proactive and contribute to achieving results (Brunetto et al., 2018). There is a considerable body of literature about the value of strong and weak ties within an organization (e.g., Granovetter, 1973; Haythornthwaite, 2002; Levin & Cross, 2004). Research shows tremendous benefits of thriving relationships within and outside your immediate team (Microsoft, 2022). It is widely accepted that interaction with colleagues is helpful for knowledge exchange, bonding, and team building. Still, it becomes more and more evident that it also influences increased well-being at work (Amorim-Ribeiro et al., 2022).

MY EXPERIENCE AS A "SAILING BUSINESS CONSULTANT"

When I left my job in the financial industry a few years ago to work remotely as an independent management consultant, this is exactly what I experienced. At first, I found it very positive that I could work very focused and without the distraction of too many meetings and other time guzzlers on my sailboat. For me, this environment is ideal for creative work and I do not even want to talk about the great work-life balance.

However, after a while—and with increasing distance to the clients—I also noticed that this form of working also has (some) disadvantages. From my point of view, the most important aspect was the more cumbersome informal

FIGURE 3.1
Focused work as a sailing digital nomad.

exchange with colleagues. Especially small questions, such as who is responsible for a certain topic or what is new in the individual teams, are often neglected. What was previously possible by simply talking to a colleague "over the desk" suddenly required a lot of additional effort and planning and tracking. What used to be done almost unconsciously on the fly in the office now required deliberate planning. This is not only my personal experience but also the conclusion of some researchers who published on this topic. They outline that disadvantages include higher demand for self-management for employees due to the reduction of interaction with colleagues and the company in general. They suggest that remote work is "effective with properly developed communication process and well-coordinated interaction of managers and remote employees" (Blumberga & Pylinskaya, 2019, p. 281). Overall, I am still convinced that remote work has many advantages and can be very efficient. Nevertheless, it requires a certain adjustment and has new implications for effective teamwork and collaboration.

CHALLENGES WITH ONBOARDING OF NEW EMPLOYEES

The above-mentioned issues are even more relevant when it comes to the onboarding of new employees. It is not easy to ensure a smooth onboarding when most operations are WFH. The onboarding process is not only about knowledge transfer, setting up infrastructure, and integrating the new employee into existing processes. It is also about factors including values and informal team rules. Usually, new employees are highly motivated, but this motivation might suffer quickly when they are in an unsatisfying WFH situation and lack integration with the organization and team. Javier Solero, for example, a manager at Google, joined the company shortly before the Covid crisis and states in the context of his onboarding experience: "I didn't have the chance to really meet and build strong connections with the people on my team" (Smith, 2022).

Probably many employees are in similar situations. It is necessary to capture tacit knowledge and transfer it into explicit (corporate) knowledge to make it accessible to all new employees (Dadgar, 2020). Some companies, like Salesforce, even AI and automation AI and automation for "some of the repetitive administrative tasks that come with every new hire—such as communicating to new employees what internal resources are available to them and what onboarding tasks need to be completed" (Spiegel, 2021). Rodrigues and his colleagues, to mention one example, conducted an exploratory study to identify key factors which have impact on productivity. Their data suggests that some factors have more impact on productivity than others (2022). They identified five factors that are especially relevant for productive WFH (Rodrigues et al., 2022):

1. Appropriate home office equipment
2. Support from management
3. Adequate place to focus on work
4. Internet infrastructure
5. Good and clear communication with colleagues and other collaborators as well as clarity in roles and responsibilities.

Still, there are many more factors with potential impact on WFH productivity and efficiency. It is also essential to facilitate feedback. Research shows that task-based learning processes and learning through interactions with supervisors and colleagues are positively related to employees' levels of work engagement. But that is not all; the strength of the relationships between these informal learning practices and the individual work engagement is related to employees' proactivity (Susomrith & Coetzer, 2019). It also must be understood that employees frequently reach out for feedback from colleagues within their department. Usually, they perceive the feedback as useful (van der Rijt et al., 2013). Adapting the management style to the new work environment and developing leadership skills for dispersed teams is crucial. Another measure is creating support teams and common developing of good practices. This includes infrastructure for knowledge sharing and communication but should also address practical questions like who (and when) to ask for advice and even allow for some casual or private chat among employees to get to know each other better.

THE "HEY JOE PRINCIPLE" – PROBLEMS AND LIMITATIONS OF INFORMAL NETWORKS

Of course, the extensive use of informal networks and "neighborly help" in the company (peer support) can also have disadvantages. In IT service management it is sometimes even referred to as the "Hey Joe principle" (CIO Editorial, 2004; Diefenthal, 2005) or the "Hey Joe effect" (Fivaz, 2017; Klassen,

2014). It is, basically, that the user is asking for help to solve a problem via an unofficial request in his/her company's internal network, and it is pointed out that the effects of the "neighborly help" can be problematic because it bypasses intended workflows; for example, the service desk as the single point of contact. This can have several negative side-effects on the service desk's performance, which is unaware of the problem. These include:

- The problem (and its solution) cannot be documented.
- There is no added value to other users seeking help, as the solution is not made available to them.
- No analysis can be made about frequency, duration, etc.
- Underlying causes cannot be identified.
- Business requirements are not prioritized and might cause conflicts with other (more important) business objectives.
- The productivity of the helping person (Joe) to perform the actual assigned and possibly more urgent and relevant tasks is reduced. This can be seen as the indirect costs of the inadequate process.

That is why it makes sense to establish knowledge management and single points of contact for specific processes and questions. The idea is to facilitate and manage the need for ad hoc collaboration when problems arise. However, the necessary infrastructure and operations need to be established and known to everyone involved before the issue becomes a problem.

THE "HYBRID PARADOX"

The "Hybrid Paradox" basically describes employees wanting the flexibility to work from anywhere but simultaneously wishing for more in-person connections (Nadella, 2021). More than 70% of employees prefer flexible remote work options; at the same time, more than 65% feel the need for more in-person contact with the team members (Microsoft, 2022). This paradox is probably challenging to solve.

There is also a shift from synchronous communication to asynchronous communication. It could be observed that employees in the WFH situation are more static and siloed, with fewer bridges between colleagues (Yang et al., 2022). Research has shown a considerable decrease in the "average monthly collaboration hours spent with cross-group ties, bridging ties, weak ties, and added ties" (Yang et al., 2022). In the long run, this might have significant adverse side-effects; for example, increased silo mentality, reduced knowledge transfer, and lost creativity. Traditional tools for social network analysis (e.g., Knoke & Yang, 2020; Scott, 2017) could be used to identify informal networks in the organization. This can help gain awareness and be a starting point for developing specific measures to manage cross-organizational communication. Technology to facilitate asynchronous communication can also help to reduce the gap.

WHAT CAN BE DONE TO IMPROVE VIRTUAL COLLABORATION?

Although the challenges are manifold, there is no reason to despair. There are many things that can be done to improve virtual collaboration. First of all it is essential to raise awareness of the challenges. It is well known that various factors are essential to remote work success. These factors include "trust in employees; embracing flexibility; strong interpersonal relationships; investment in quality tools and technology; hiring for culture fit; culture-focused mentorship; strategic in-person encounters; and continuous culture improvement" (English, 2022). Angela Ashenden, an analyst at CCS Insight, gives the following advice: "It's about changing the culture, the way people work together, and the way teams are organized, so they operate in a cross-company fashion rather than in little departmental pockets – and so the cultural mindset has to change, not just the tools" (Everett, 2022, p. 22).

In particular, these challenges are sometimes overlooked with the rush toward "emergency WFH setups," as we have seen a lot during the COVID-19 pandemic. While the focus is (possibly for good reasons) to ensure the core business processes, it might be the case that the informal networks and informal processes are somewhat neglected during this change. Sometimes this is not immediately an issue because the employees that used to work in office settings with their colleagues have already established informal networks and corresponding team rules. It looks pretty likely that they do not change these established routines just because they are in WFH mode from one day to another. This is probably not the worst strategy in the case of the Covid crisis but things work differently in the WFH environments with henceforth virtual teams. About 59% of all hybrid employees and about 56% of remote employees stated that they have fewer work "friendships" since the shift to hybrid or remote work (Microsoft, 2022). Many employees feel lonely in the WFH situation (Hadley, 2021). Therefore, a change is needed in the medium and long term to facilitate bonding in the organizations.

Actually, in this situation, we can learn from the experience of large organizations that have already had operative virtual and distributed teams for quite some time in various contexts. Some McKinsey consultants share the following advice:

> Just as formal hierarchical structures define management roles, formal network structures define collaborative professional ones. In this way, such networks can enable large companies to overcome the problems of very large numbers by creating small, focused communities of interest integrated within larger, more wide-ranging communities.
>
> Bryan et al., 2007, p. 86

A practical implication could be that there are new roles within the company and team structures to "manage" the collaboration and communication among the employees. Sharing information with a team member across a desk is not too difficult, but most employees today do not sit at a desk in the same office building. It seems to be important to care about informal communication and informal professional networks because they are positively associated with improved organizational culture (Brunetto et al., 2018). Still, this is not necessarily an easy task. Informal networks may not be visible to management and, even when they are, it may be difficult for organizations to take full advantage of them. "Unintended barriers, corporate politics, and simple neglect can keep natural networks from flourishing. At worst, informal networks can make dysfunctional organizations even more so by adding complexity, muddling roles, or increasing the intensity of corporate politics" (Bryan et al., 2007, p. 83).

More than ever, it is essential for the manager to set up regular one-to-one meetings with employees. This should not be confused with team meetings or other meetings because these types of meetings should be short and stringent. Excessive palaver meetings with many participants are usually neither efficient nor pleasant for the people involved. The manager's role is to support employees and ensure a productive work environment. There are many factors that a manager needs to consider, such as the health and safety of employees and those around them, personal circumstances, knowledge, infrastructure, workload, and their ability to complete their work. Managers are ill-advised when they believe returning to the office is the only solution for rebuilding the social capital which might have been lost over the past two years. They should prioritize relationship-building time, knowing remote and newly onboarded employees will need additional and different support.

Managers play an essential role in fostering close team bonds and acting as links in helping employees build and broaden their networks within the organization (Microsoft, 2022). Constance Hadley, an organizational psychologist, also suggests rethinking the performance-management systems. Organizations may overall profit when they increase the benefits and reduce the risks associated with employees for reaching out to coworkers. This means it is necessary to notice and reward employees for making the first move and responding supportively to colleagues' needs and questions (Hadley, 2021).

Still, it is not only about the line manager but also the interaction with other colleagues. Employees with thriving relationships report better overall well-being and feel more productive at work. They are generally more satisfied with the employer and have less intention to change the employer compared to employees with struggling relationships. This applies to the direct team and also to relationships in the broader organization (Microsoft, 2022). Some companies support this with a buddy or mentorship system within the organization. The onboarding buddy helps the new employee to understand the workplace, structures, and processes. This buddy can also give advice and provide context. Generally, buddy systems can help the new employees' satisfaction and productivity (Cooper & Wight, 2014; Klinghoffer et al., 2019).

For productive remote work it is important for employees to understand that remote work "is a skill anyone can acquire, rather than something for which certain kinds of people are either well or poorly-suited" (Rodrigues et al., 2022, p. 30). This means that organizations should cultivate adaptive mindsets and encourage employees to develop growth mindsets about working remotely (Rodrigues et al., 2022). Another area that requires some attention is the management of cross-group ties and bridging ties in the organization. An onboarding buddy should provide an opportunity to demonstrate and develop leadership skills and should appreciate new employees in the organization. This is also a question of corporate culture and valuing the commitment of individuals.

SUMMARY

Information networks can be an important pillar in an organization. They are not only helpful for knowledge transfer but also for getting things done and increasing rapport. However, it should not be a replacement for sound business processes and proper corporate communication strategies. In general, remote work causes the collaboration and informal network of employees to become more static and siloed. There are fewer bridges between disparate teams. All in all, these effects tend to make it more difficult for employees to acquire and share new knowledge and information across the organization (Yang et al., 2022). Line managers are crucial in promoting team communication and establishing links with and among employees. It is a skill that needs to be cultivated in an organization and should become part of the "corporate DNA." Still, it is also vital to encourage employees to make proper use of networking strategies and build up their ties within the organization (and beyond).

REFERENCES

Amorim-Ribeiro, E. M. B., Neiva, E. R., Macambira, M. O., & Martins, L. F. (2022). Well-being at work in processes of organizational change: The role of informal social networks. *RAM. Revista de Administração Mackenzie, 23*(1), eRAMG220125. https://doi.org/10.1590/1678-6971/eramg220125

Awada, M., Lucas, G., Becerik-Gerber, B., & Roll, S. (2021). Working from home during the COVID-19 pandemic: Impact on office worker productivity and work experience. *Work, 69*(4), 1171–1189. https://doi.org/10.3233/WOR-210301

Bergmann, F. (1994). New work, new culture (S. van Gelder, Interviewer) [In Context, IC#37 It's About Time!]. www.context.org/iclib/ic37/bergmann/

Blumberga, S., & Pylinskaya, T. (2019). Remote work advantages and disadvantages on the example in IT organisation. 275–282. www.nordsci.org/nordsci-library/p/remote-work-%E2%80%93-advantages-and-disadvantages-on-the-example-in-it-organisation

Brunetto, Y., Xerri, M., Farr-Wharton, B., & Nelson, S. (2018). The importance of informal professional networks in developing a proactive organizational culture: A public value perspective. *Public Money & Management, 38*(3), 203–212.

Bryan, L. L., Matson, E., & Weiss, L. M. (2007). Harnessing the power of informal employee networks. *Electric Perspectives, 32*(6), 82–86.

CIO Editorial: ITIL vereitelt "Hey-Joe-Prinzip." (2004, August 11). *CIO*. www.cio.de/a/itil-vereitelt-hey-joe-prinzip,802337

Cooper, J., & Wight, J. (2014, October 26). *Implementing a buddy system in the workplace.* PMI® Global Congress 2014, Phoenix, AZ. www.pmi.org/learning/library/implementing-buddy-system-workplace-9376#:~:text=What%20is%20a%20Buddy%20System,or%20months%20on%20the%20job.

Dadgar, M. (2020). In T. Endress (Ed.), *Digital Project Practice: Managing Innovation and Change* (pp. 78–88). Tredition.

Diefenthal, A. (2005, August 23). Das "Hey-Joe"–Prinzip. *Midrange*. https://midrange.de/das-hey-joe-prinzip/

English, L. (2022). Centric consulting case study: Culture is the key to remote work success. *Work, 71*(2), 295–298. https://doi.org/10.3233/WOR-210701

Everett, C. (2022, January 18). Ensuring hybrid work works in the new normal. *Computer Weekly*, 19–22.

Fivaz, A. (2017, May). *Der Hey-Joe-Effekt – verdeckte Kosten in der Informatik.* https://bsg.ch/publikationen/der-hey-joe-effekt-verdeckte-kosten-in-der-informatik/

Granovetter, M. S. (1973). The strength of weak ties. *American Journal of Sociology, 78*(6), 1360–1380.

Hadley, C. N. (2021). Employees are lonelier than ever. Here's how employers can help. *Harvard Business Review.* https://hbr.org/2021/06/employees-are-lonelier-than-ever-heres-how-employers-can-help

Haythornthwaite, C. (2002). Strong, weak, and latent ties and the impact of new media. *The Information Society, 18*(5), 385–401. https://doi.org/10.1080/01972240290108195

Ipsen, C., van Veldhoven, M., Kirchner, K., & Hansen, J. P. (2021). Six key advantages and disadvantages of working from home in Europe during COVID-19. *International Journal of Environmental Research and Public Health, 18*(4), 1826. https://doi.org/10.3390/ijerph18041826

Kienbaum, F. (2017, June 22). New work pulse check 2017. *Kienbaum Blog.* www.kienbaum.com/de/blog/new-work-pulse-check/

Klassen, M. (2014, January). Run SAP like a factory. *Galileo Group: Lessons Learned Article.* www.galileo-group.de/files/en/why-galileo/news-items/Case%20study%202015-001%20Mettler%20Toledo.pdf

Klinghoffer, D., Young, C., & Haspas, D. (2019). Every new employee needs an onboarding "buddy." *Harvard Business Review.* https://hbr.org/2019/06/every-new-employee-needs-an-onboarding-buddy

Knoke, D., & Yang, S. (2020). *Social network analysis* (3rd edition). SAGE.

Leonardi, P. M. (2021). COVID-19 and the new technologies of organizing: Digital exhaust, digital footprints, and artificial intelligence in the wake of remote work. *Journal of Management Studies, 58*(1), 249–253. https://doi.org/10.1111/joms.12648

Levin, D. Z., & Cross, R. (2004). The strength of weak ties you can trust: The mediating role of trust in effective knowledge transfer. *Management Science, 50*(11), 1477–1490. https://doi.org/10.1287/mnsc.1030.0136

Microsoft. (2022). *2022 Work Trend Index: Annual Report.* https://ms-worklab.azureedge.net/files/reports/2022/pdf/2022_Work_Trend_Index_Annual_Report.pdf

Morikawa, M. (2022). Work-from-home productivity during the COVID-19 pandemic: Evidence from Japan. *Economic Inquiry, 60*(2), 508–527. https://doi.org/10.1111/ecin.13056

Nadella, S. (2021, May 21). *The hybrid work paradox.* www.linkedin.com/pulse/hybrid-work-paradox-satya-nadella/

Rodrigues, E. A., Rampasso, I. S., Serafim, M. P., Filho, W. L., & Anholon, R. (2022). Productivity analysis in work from home modality: An exploratory study considering an emerging country scenario in the COVID-19 context. *Work, 72*(1), 39–48. https://doi.org/10.3233/WOR-211212

Sarkar, S., & Kedas, S. (2022). Globally distributed talent communities: A typology of innovation problems and talent characteristics. *Thunderbird International Business Review*, tie.22272. https://doi.org/10.1002/tie.22272

Scott, J. (2017). *Social network analysis* (4th edition). SAGE.

Simenenko, O., & Lentjushenkova, O. (2021, September 16). Advantages and Disadvantages of Distance Working. 18th International Conference: Perspectives of Business and Entrepreneurship Development, Brno, Czech Republic.

Smith, M. (2022, April 8). The 3 "biggest mistakes" companies make with return to office, according to Google's head of Workspace. *CNBC Make It*. www.cnbc.com/2022/04/08/google-workspace-vp-the-3-biggest-mistakes-companies-make-with-return-to-office.html

Spiegel, S. (2021, April 28). The future of work at Salesforce: Digital, human and connected. *News & Insights*. www.salesforce.com/news/stories/salesforce-future-of-work/

Susomrith, P., & Coetzer, A. (2019). Effects of informal learning on work engagement. *Personnel Review*, *48*(7), 1886–1902.

van der Rijt, J., den Bossche, P. V., & Segers, M. S. R. (2013). Understanding informal feedback seeking in the workplace: The impact of the position in the organizational hierarchy. *European Journal of Training & Development*, *37*(1), 72–85.

Yang, L., Holtz, D., Jaffe, S., Suri, S., Sinha, S., Weston, J., Joyce, C., Shah, N., Sherman, K., Hecht, B., & Teevan, J. (2022). The effects of remote work on collaboration among information workers. *Nature Human Behaviour*, *6*(1), 43–54. https://doi.org/10.1038/s41562-021-01196-4

Zajac, S., Randall, J., & Holladay, C. (2022). Promoting virtual, informal learning now to thrive in a post-pandemic world. *Business and Society Review*, *127*(S1), 283–298. https://doi.org/10.1111/basr.12260

4

Why New Work Is the Result of Fundamental Changes in a Company*

Philipp Lehmkuhl
MEWA Textil-Management, Wiesbaden, Germany

CONTENTS

* Translation by: Dr. Tobias Endress

DOI: 10.1201/9781003371397-4

THE CULTURE SHOULD CHANGE – DO SOMETHING!

Many people responsible for digital transformation are given the task of driving not only digitization but also changing the culture—more agility, more speed, and a climate that is generally conducive to innovation should prevail in the company. But what actually is culture? My insight is that culture is a diffuse collective term that initially hides behind other collective terms. It is about values and tradition, family businesses, the spirit of the founders from days gone by, and an actual or perceived mission. Sometimes, however, it is also about the vision and purpose of the company, rarely about structure and processes. For me, the culture behind all these terms is primarily how the company is really set up internally. So culture is a causal consequence of how the company is set up structurally, process-wise, and technologically. You will not be able to change the culture without working at the root. New Work—that is, a new way of working—is, therefore, a topic that starts at the root and then, in its manifestation, brings to light a tangible culture that makes the company agile, strong against disruptions, strong on the recruiting market, and, most importantly, strong with the customer.

What Do I Call "New Work"?

"New Work" is a term often used without a clear distinction and it appears that there is no clear and a uniform understanding in business life. That is why I write: What do *I* call New Work, and not "someone else." There is no fixed agreement in the business community on a particular definition, as is so often the case with the new terms of digital transformation.

In the end, several factors come together that can possibly be called "New Work." First of all, it is described as something new, which is distinct from the classic, established way of business organization and collaborative work is arranged. Unfortunately, partial aspects of the overall picture are often emphasized in such a way that one might assume that this is now "New Work." I will try to put this into an overall context. It remains to be seen whether this is really so new because many companies already work successfully according to this culture. It should be mentioned that I am only reporting based on experiences I have made in the context of digital transformations in various companies.

New Work is a term that is supposed to describe a different culture in a company, and this is the result of a combination of these essential changes:

1. Customer centricity
2. Structure of a company
3. Management methodology
4. Agility (as a result)

New Work is now becoming interesting for well-established companies in the traditional economy for two different reasons: First, the companies that operate under this model are above average in success (see e.g., Altindag & Siller, 2014; Andriyanty et al., 2021; Wanto, 2021); this is putting established companies under increasing pressure to move, as they run the risk of being disrupted by competitors or companies from outside their industry or startups that are above average in success if they operate beyond their ambitions. And that is a trend we are increasingly seeing. Although it might be worth mentioning that "it can be shown that New Work does not automatically lead to higher firm performance" (Block, 2019). It needs careful but complete implementation in the organizational context. New Work can only succeed if the structural and technological conditions are in place.

Second, a rising generation of young employees no longer wants to spend their working time according to the principle of "time for money," but to be able to understand the meaning and purpose of the task and to feel the effectiveness of their own efforts. The individual employee wants to see a visible contribution to their work. Therefore, the individual employee must feel more efficient in their work. This means that they must also be able to take more responsibility for their work and thus have more discretionary power of their own. Basically, this is the "entrepreneurial thinking in the employee", which is often desired but often not found in reality. However, to make the right decisions, all employees must be focused on one goal, which must be coordinated, staggered and structured throughout the company. This automatically leads to a weakening of the tasks and power of the classic hierarchy, the classic middle management. And here we are with the New Working methods, which in fact do not require a classical pyramid hierarchy and different middle management—or none at all. Why this is the case is shown in the following section.

Customer Centricity – What Does New Work Involve?

In summary, New Work can be described as teams developing products, projects, solutions, and processes on their own responsibility, across divisions and disciplines, within a set timeframe. This is done within the team's own

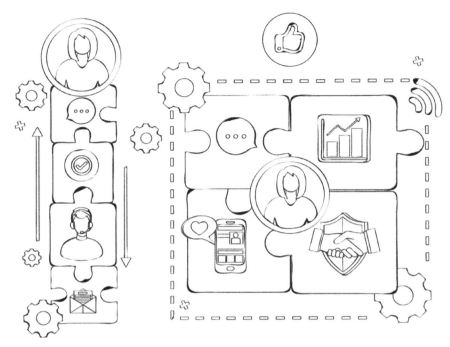

FIGURE 4.1
The customer is in the center of our minds and hearts and processes (vertical silos to be flipped horizontally).

role-based hierarchy and according to an agile working methodology. The whole process is organized via a management framework OKR (objective key results) in many teams in parallel and is coordinated across the board.

This sounds simple but it also requires some preconditions—and they are pretty demanding. To be clear: remote working, virtual working, and fancy interiors are not New Work—these are only manifestations that are possible but not mandatory. I will provide more details on this later.

Prerequisites for New Work

These self-organized cross-functional teams must be empowered to make decisions. This means that the person taking on the role of responsibility must not be controlled by interventions from the line organization. Otherwise, the flow of work is impeded and, without knowledge of the actual situation, decisions are pushed into development that discourages the entire team and usually does not lead to better results. Teams must work directly with the customer or user and in a methodically disciplined manner. Only by working with the person whose problem is actually being solved does a customer-centric solution emerge.

These teams must understand and holistically implement the agile work methodology. Let us say agile freestyle is the death of the subject. The technical prerequisites must be in place to be able to work agilely. This includes cloud-based development, a DevOps development team, and a scalable service infrastructure like you find in hyperscalers (Hoogendoorn, 2021), among other things.

But to make that happen the team needs to know exactly what it needs to accomplish and what its goal is. This requires a detailed declination of the OKR target framework in the enterprise.

So What Needs to Be Done in Concrete Terms?

As it does so often, it starts at the very beginning with the principles, a cleanly and aptly formulated vision, mission, and strategy. These guidelines are prepared in an emotional and customer-centric way so that they give their employees a direction as to what purpose their daily work serves and, at the same time, provide the compass for decisions within teams. Because a team that has understood together in which direction to pull the strings is faster, stronger, and more effective, proving to be the cornerstone for cultural change and thus for New Work.

The most important aspect is a term that is often overlooked because it is taken for granted. Many companies have moved away from it in their self-purpose and self-centeredness, in their outdated silo structure, and out of the comfort of their own management and employees: "Customer centricity." What does it mean?

Structure of a Company

Customer centricity means that all processes in the company must be thought of from the customer or user perspective. Whether it is business-to-business or business-to-customer, everything turns irrelevant. In practical terms, this means that a process always starts with the user, continues throughout the organization to the suppliers/manufacturers/suppliers, and back to the user or customer. It is a closed loop. If this is the case, one ensures that all processes and thus the business model in the company are aligned end to end with the customer.

The existing processes, therefore, reflect the degree to which the business model is customer-centric. Companies that are more customer-centric and can sustainably satisfy their customer needs better than competitors in the industry gather meaningful data about market changes faster than companies that revolve around themselves and are thus superior to other companies.

The next step is to check whether the process-based implementation of the business model really focuses on the customer. The processes must now be analyzed end to end. In the course of this, the customer lifetime value (CLV) must be established as the measure of all things, and all process steps must be checked to see whether they increase the customer lifetime value. Data supply chains, process gaps, technology gaps, and other frictions must be identified and subsequently addressed in the implementation of the strategy so that digital customer-centric end-to-end processes are enabled to increase CLV (Groeger & Buttle, 2015).

And here comes the next paradigm shift. Traditional companies are driven by metrics. However, metrics are an internal view, not the customers. A customer is not interested in a company's metrics. Key figures are management and shareholder view. Therefore, KPI (key performance indicator) systems are not suitable as a controlling goal-setting instrument. Furthermore, employees are certainly not emotionally responsive to key figures, but via purpose and vision. Therefore, KPI target systems are generally not suitable for building customer-centric organizations.

Remark

Do not get me wrong, this does not mean that there should not be KPI schemes, only that these should not be given as a target or goal-oriented system, but only as a controlling aspect of a company. Of course, attention must be given to ensuring that liquidity and earnings are in place. But these are much easier to achieve if the entire team pulls together end to end in the direction of the market and the customer, friction losses are eliminated, the energy is focused on one goal, and every single employee knows what he/she is working and fighting for every day. It is then much easier to deprioritize topics, check the objectives of individual profiling artists in the company regarding the customer, and calculate the value contribution in CLV. What enormous efficiency and effectiveness gains can a company achieve with this?

Suppose the goal of the company is that in an end-to-end process, a starting process segment must selfishly pay attention to its goal achievement at the expense of a subsequent process segment. In that case, a continuous process cannot run efficiently. Often, areas or departments are responsible for a specific process segment. Each process segment (i.e., each department) will therefore align and optimize itself according to its goal. Thus, the local optimum prevails and not the continuous overarching optimum for the entire company or the customer. Therefore, a process-based target system is required, which is oriented to an end-to-end process, which can only have one orientation in the overall process: the customer focus and thus the CLV. This automatically creates an overarching end-to-end process with a uniform orientation and a logic for the respective process sections that are automatically derived from it. As a result, a company no longer has a classic divisional organization but only a process organization with a pure cross-functional agile team structure. That includes an end-to-end process owner with the power comparable to a board member to really drive the process toward customer centricity without forgetting the necessary company metrics.

This is a process-based structure, the real change and the basis for any agility and new work. This is the foundation for the "transformation" in culture—this is the word "transformation" in the term "digital transformation." In order to ensure the professional excellence of the team members, excellence groups are formed in which the professional-specific exchange and further training are guaranteed. And there we are again at the basic idea of New Work. Agile cross-functional collaboration.

Management Methodology – Objective Key Results

A target system that is better suited for the implementation of New Work than a key figure system is OKR (Doerr, 2018). This methodology from the 1970s became world-famous through Google's adaptation at the turn of the millennium and has proven to be precisely an efficient tool for customer-centric companies.

OKR is essentially the tool to "agilize" the entire company. It allows first to reformulate vision and mission into goals in such a way that the implementation of the strategy can be realized in medium- and short-term goals and is always in the same temporary units (sprints). OKR helps to provide the goals with concrete customer benefit-based metrics so that the individual teams know specifically what their projects contribute to and how they act in the overall context. At the same time, the methodology of OKR gives a permanent progress status, as well as adjustment possibilities. Only from this top-down planning does a complete picture emerge for the implementation of an overall strategy that can run for several years and still allow enough room for adaptation to change circumstances through its agile approach. It sounds like a waterfall but it is basically the enterprise-wide agile implementation of the backlog, which is fed by the enterprise's strategy.

OKR, like all agile methods, is managed with disciplined governance so that from the smallest cog to the big wheel all teams work together like greased clockwork. This goal system always works in a customer-centric way. Therefore, customer-centric end-to-end processes are also the key to success. After all, customers are not interested in internal processes but rather in the best price-performance ratio.

New Work, therefore, describes the target state of agilization of the entire company. OKR enables this agilization of a company as the whole in its entirety if the procedural, structural, and technical prerequisites for it are created.

Agility (as a Result)

If the points mentioned above—centricity, structure of a company, and management methodology—are actually implemented in this way, agility is no longer something you have to introduce; it is there as a result. It is, as you might imagine, really not easy to bring about such a change. Many companies do not even want it—they are doing well, and why try so hard, as you cannot really prove that things will be much better afterward. So the disease is not cured, but only the symptoms are treated and, for the time being, it seems as if something is happening. Top management can then claim that something is being done but that it is a long process and that no one will be interested in five years anyway. So now we look at the symptoms, why fancy offices are related to the cloud, and subsequently at the resistance in the companies.

Symptoms

Why is New Work often equated with open-plan offices, with more playful furnishings? What sense does that make anyway?

Open-plan offices with fancy furnishings, corners where you can talk, and corners where you are supposed to work quietly, and various playful things like swings, etc., have become established as a stylistic device in many companies, driven by numerous architectural firms and facility management. This is only a symptom or an expression without a connection to the reason behind it. This expression often is flaunted as a fig leaf, just like excessive use of undefined buzzwords, but behind it everything is the same.

So, What Connects Fancy Open-Plan Offices with the Cloud?

One of the principles of New Work is that agile teams are temporarily assembled individually according to product or problem. This works particularly well if you do not have individual offices, each occupied by a few employees, but instead have more flexible areas in which teams can come together according to the relevant constellation.

And this does not mean meeting rooms but rather rooms and areas in which a project has a home on a continuous basis and to which the respective responsible persons temporarily switch. This is a big difference from a meeting room with its fixed seating arrangement, which can be booked by the hour. Here, the spatial structure of the meeting room provides the framework for the project and its members and not vice versa. It is actually amazing that we let a few chairs, tables, and a beamer determine the way we work.

Now Let Us Take it Down a Notch

If necessary, this agile team also includes a DevOps part. As the name DevOps suggests, this is an amalgamation of at least two topics or teams, namely development and operations. These two topics should actually be kept separate for IT security reasons but in the agile world of work they act very closely and yet safely together, as long as the separation of responsibility (segregation of duty) is adequately observed. This is where the idea comes from acting persons who can see and discuss among each other in the morning and subsequently exchange information over short distances during the day. For example, the product owner, business process owner, agility master exchanges information with the developers and operations, etc., if there are queries or changes. In the course of this, the open-plan offices with short paths, where people meet together in the morning daily, can quickly exchange information with the responsible person at any time, make a lot of sense with the tremendous flexibility and application purpose-focused setup.

DevOps is there to ensure, together with the conceptual part, a really continuous ongoing development and deployment process and to provide a flexible implementation of the conception in flow together with the customer or user. This is achieved by setting up the development environment in the cloud. In addition, web services and a range of SaaS solutions, agile project management tools, and a CI/CD (continuous integration/continuous delivery) pipeline are used. The real agility of the DevOps team is shown only if it can react quickly and flexibly to changes and new requirements without first having to procure infrastructure and set up the application landscape at great expense. What use is it to the customer-facing product owner if they have to wait six weeks until the IT procurement department has ordered a new server and installed a specific application? In this respect, the cloud is a mandatory prerequisite for actual agile and fast customer-focused work. So cloud is related to spatial collaboration via DevOps and the agile customer-involving team, and thus also to the fancy chic flexible open-plan offices.

Now, I hope you can understand how fancy open-plan offices with great furniture are related to the cloud. Unfortunately, too often in many companies, only the office is spruced up accordingly, but all the rest is not. This counteracts the topic of New Work and burns the chances because there will be no effect, except for complaints from employees looking for their workplace and wanting to have their peace when they want to pursue their regular activities in the department. Why am I not allowed to put my photo of my children here anymore? The legitimate question remains unfounded if you do not create the working methods and structures behind it. This leads to a lack of understanding and frustration, and means that a company is gambling away the potential associated with New Work.

IS VIDEO CONFERENCING, HOME OFFICE, WORKCATION, AND WORKING OUT OF BALI OF THE GEN-Z NOT ACTUALLY "NEW WORK"?

Video conferencing, the home office, and decentralized cross-continental working only replace the fancy open-plan office, but only if the agile working methods have been introduced accordingly. In the meantime, it has been proven that these methods also work exclusively virtually. In this respect, virtual working is utterly compatible with agile working and a veritable component of New Work. The question even arises whether the fancy open-plan offices in the future can probably be more of a space for identity formation in the spirit of the company vision for a team than the necessary workspace. I venture a prediction here that in the future we will have pop-up offices in various cities around the world near the homes of team members, where temporary teams come together to work on a problem solution. In addition, we will see more project workers who temporarily feel a sense of belonging to a project or task, more than to the company itself that finances the project. Only a company that has set up its internal structures and target system as described above will be able to set up ad hoc, target-based teams worldwide that carry out their projects in a decentralized manner. This will be a crucial capability in the future to keep up with the dynamics.

The War for Talent is Real

Unfortunately, in the wake of digitization, we will have a trend for years to come in which we have to transform while maintaining the old processes and methods digitally. This phase will possibly continue until 2040, until AI and robotics systems are so mature that they replace human work across the board and the potential of the digital revolution is actually seen tangibly in the labor market as well. In this phase we will feel a shortage of skilled workers in almost all areas, and especially in all areas of digitization and areas related to New Work. In this phase young professionals will find themselves in a land of milk and honey because they will be able to demand high salaries and special conditions, such as permanent remote work while starting with a low output. And the companies will pay for it, leading to a lack of understanding among long-established employees. This will be a new task for HR to manage this balancing act.

New Work will show its dark side here: This upcoming generation will be the one that will not reap the fruits they have worked on with full force, but will feel them. The career that started so easily and glamorously professionally will later become a tough competition, and only those will then remain who have secured a place in the company through an above-average understanding of the interrelationships of the new world and a high degree of appropriation of the company's vision to lead the company's fortunes in the future AI- and data-base.

Why Does New Work Fail in Traditional Established Companies?

New Work fails in companies because the supposed pillars of the company have the most to lose and therefore actively work against structural change. This refers to the pyramidal middle management, because this is actually no longer necessary in New Work. It is replaced by role-specific management.

With the introduction of role-specific management, traditional middle management loses its right to decide things.

This is why the greatest resistance in a company can be found here. Middle management employees from their mid-40s upwards have the most to lose. It is only human and understandable because they all have their house to pay off, the car loan, the daughter or the son may still be studying, and they are basically already on the verge of moving toward pension. Destruction, real change, really working again, getting back into the fold, and taking on a special task instead of just flying over it is incredibly difficult for the vast majority of executives; they have worked too long to reach the desired position, leaving them no choice but to defend their garden fence with all their might. I understand that, but it is still wrong. These garden fences must be broken down! The task for middle management must basically disappear, otherwise there is no real cultural change. Somebody has to bear the burn of this cultural change, and it seems to be this middle-aged generation of working people.

And that is where the problem lies: How does a traditional company work? The classic company operates in such a way that the supervisory board relies on the management board, the management board relies on its first management level, the latter relies on the second management level, and so on. Now it takes a board of directors and the supervisory board that has understood the new methods with their advantages to bring about such a change. Any board that is not fully convinced and driving this change personally will, in case of any doubt, rely on its executives and ask them what they think of a possible change in this direction. Since the executives have the most to lose, they will find all kinds of arguments to defend their garden fence, and they are surprisingly united on this. In the end, the board is uncertain whether it can possibly introduce this change at all if its executives are against it. It will decide against it in case of doubt and recommend this to the supervisory board. That is the end of structural change in an established company, and thus of New Work, as long as the numbers are right.

What Will Be the Role of Middle Management in the Future?

Middle management must focus on the task of people management and personnel development. Here, too, there is a basic problem. Today, managers are often those who have distinguished themselves through diligence, expertise, sympathy, or by playing hardball in the finger-wagging of the distribution of competencies, not through special people management skills. But this is exactly what is becoming more and more critical. The task is to lead the excellence groups, also known as guilds, which ensures that the employees deployed in the cross-functional teams exchange professional knowledge with their peers and develop further; the personal development of each individual in line with the overall strategy becomes the task of middle management. In the process, they become service providers and coaches for the employees. There is no longer any budget sovereignty, no technical project decisions, no headcount, and no divisional responsibility for a process; instead, they serve the personal progress of the team. Of course, middle management can also take on role functions such as process owner, OKR manager, product owner, and other roles; this is up to individual preference.

HOW DO YOU START IN A COMPANY THAT DOES NOT WANT TO GO ALL OUT RIGHT AWAY?

You start with pilots in small, delimited topic blocks. In order to get to know the new agile methods and gain initial experience, train employees, and build up expertise, it is advisable to start with the first self-organized cross-functional teams in delimited subject areas with as few internal processes intersections as possible. These projects should be accompanied by an experienced agility coach to ensure that the methodology is implemented properly and the discipline required to work in an agile manner is maintained. These pilots should be accompanied by top management with a high level of attention so that both sides can assess the methodology and also each other's expectations. It is often wrongly assumed by conventional management that agile projects can be planned from A to Z, similar to the classic waterfall project. The deceptive security of waterfall planning on paper suggests the ability to implement a project according to the planning. However, all experience shows that this is hardly possible in any company and certainly not for implementing an overall strategy. Therefore, it is imperative to learn agile skills.

These pilots should also give top management the necessary confidence in the new methods because it requires much more discipline to work cleanly in a team according to roles, within the methodology, and to pursue the goals in an agile manner on a daily basis. It should not lead on the way, such as the case in a classic project, which is even carried out across several areas, in addition to the regular working day. On the other hand, agile teams usually reach their goals faster and are more customer- and user-oriented.

SUMMARY

In summary, New Work results from fundamental changes in a company. Many startups can set themselves up directly in such a way that an end-to-end process structure and agile working organization is created. For traditional companies, this transformation will be a tour de force that, once implemented, will enable the benefits of New Work in all its facets. It requires careful implementation in the organizational context. Top management and the board need to steer the path for these transformation efforts. Targets and strategies to achieve them need to be developed and implemented. New Work can only succeed if the relevant preconditions, in terms of structural and technological infrastructure, are in place.

REFERENCES

Altindag, E. A., & Siller, F. (2014). Effects of flexible working method on employee performance: An empirical study in Turkey. *Business and Economics Journal, 05*(03). https://doi.org/10.4172/2151-6219.1000104

Andriyanty, R., Komalasari, F., & Rambe, D. (2021). The effect of work from home on corporate culture mediated by motivation, work behavior, and performance. *Jurnal Aplikasi Manajemen, 19*(3), 522–534. https://doi.org/10.21776/ub.jam.2021.019.03.06

Block, C. (2019). *Top Management und New Work: Eine empirische Untersuchung zum Einfluss von Strategic Leadership auf Organisationen in der neuen Arbeitswelt* [Universität St.Gallen]. www.e-helvetica.nb.admin.ch/api/download/urn%3Anbn%3Ach%3Abel-1412247%3ADis4874.pdf/Dis4874.pdf

Doerr, J. (2018). *Measure what matters: How Google, Bono, and the Gates Foundation rock the world with OKRs.* Portfolio/Penguin.

Groeger, L., & Buttle, F. (2015). Customer lifetime value. In C. L. Cooper (Ed.), *Wiley encyclopedia of management* (pp. 1–3). John Wiley & Sons, Ltd. https://doi.org/10.1002/9781118785317.weom090070

Hoogendoorn, I. (2021). Public cloud integration. In *Multi-site network and security services with NSX-T* (pp. 303–316). Apress. https://doi.org/10.1007/978-1-4842-7083-7_10

Wanto, H. S. (2021). The correlation of organizational culture, organizational learning, information technology, competitive strategy and company performance: A review. *Russian Journal of Agricultural and Socio-Economic Sciences, 115*(7), 138–149. https://doi.org/10.18551/rjoas.2021-07.15

5

New Work and Collaborative Cheating – Lessons from the VW Emission Scandal

Christopher M. Castille
Nicholls State University, Thibodaux, Louisiana, USA

Tobias Endress
Asian Institute of Technology (AIT), Bangkok, Thailand

CONTENTS

BUSINESS ETHICS AND THE "OLD WORK" CORPORATE CULTURE

Bergmann's (2019) vision of a New Work culture calls organizational hierarchies into question, emphasizes creating a purpose-driven (rather than a purely financially-driven) economy where workers come together, develop shared work identities that work from common goals toward bettering our society, and jointly pursue deeper meanings of their work. This vision stands in stark contrast to the "Old Work" culture, which emphasizes such things as a bottom-line mentality and prioritizes short-term increases in financial outcome (e.g., stock prices) above the concerns of other stakeholders (e.g., operating transparently, long-term stakeholder value). We see much merit to this vision in that work arrangements can serve deeper psychological and social needs (e.g., doing exciting work in a collaborative manner, focusing attention on tasks that positively impact others, helping workers find and build a sense of community and find a deeper purpose to their work).

Digital transformation has become one of the most prominent buzzwords of recent times, and it is one of the significant trends in business and society. However, digital transformation also implies new challenges, such as consumer and employee responses to the use of AI at work, and ethical questions, such as how to transparently deploy AI in organizational decision-making structure (PricewaterhouseCoopers, 2020). Processes and system architectures become increasingly complex, but to the main users of the systems (e.g., managers, employees), they are increasingly opaque. In other words, as digital transformation complicates the systems within which we all work, it can be difficult to understand precisely how some new technology (e.g., AI) works to the benefit of key stakeholders (e.g., managers, employees, customers, government authorities, society at large).

Not surprisingly, business leaders bowing to performance pressures (e.g., VW management, see Ewing, 2017), falling down on some ethical slippery slope (see Welsh et al., 2015), or succumbing to policies and practices that promote ethical blindness (Martin et al., 2014) can use the opacity of new technologies to gain an unfair advantage over rivals in the marketplace (i.e., cheat). One case where this is abundantly clear is the VW emission scandal. Briefly, the VW emissions scandal involved a decade-long collaboration between managers and software engineers across various companies (e.g., Audi, Robert Bosch, VW) to build and maintain software (i.e., a "defeat device") that would help VW to defeat routine emissions testing in the European Union (EU) and United States. All vehicle types undergo emissions testing prior to being sold in the marketplace, one purpose of which is to minimize adverse impacts of burning fossil fuels (e.g., control the amount of CO_2 in the atmosphere; control the amount of asthma-irritating and cancer-causing particulate matter). By defeating emissions testing, automakers such as VW can operate rather profitably, particularly if their deceptions go undetected, which is why history is littered with defeat devices created by players within the automotive industry. Indeed, several well-known automakers have created a defeat device at some point in their history (e.g., Chrysler, Ford, General Motors, Honda, Toyota, Volvo Trucks, and BMW).

As this book highlights the value of New Work philosophy for digital project practice, we use the VW emissions scandal for two reasons. First, the vision of New Work brought forth by Bergmann (2019) glorifies certain aspects of the work experience to which the VW emissions scandal reveals a dark side: Identifying with some broader entity (e.g., organization, manager, work team, or broader cause) and collaborating with others over common goals are fundamentally human activities that can have bad consequences. Although VW is clearly an organization characterized by an "Old Work" culture (i.e., rigid hierarchies, authoritarian leadership), it reveals the dark side to both identity (see Conroy et al., 2017) and collaboration (see Briggs et al., 2013). Managers of digital projects would be wise to take lessons from VW's mismanagement, such as adopting a values-based perspective to business decisions (see Mansouri, 2016), which we explain in greater detail near the end of this chapter. Second, we wish to exemplify how, lacking a values-based perspective, an opaque system can be co-opted by leaders who, bowing to

FIGURE 5.1
Old work: high pressure/toxic work culture/finger-pointing.

pressures to perform, may collaboratively cheat. Indeed, in the case of VW, the case can be made that executives knowingly facilitated or nurtured such an outcome.

We submit that New Work is more than working in the home office or a nicely decorated office space. New Work refers to a work arrangement where the pursuit of meaning in life is worth honoring; ergo, good business ethics are essential to New Work arrangements. New Work stands in contrast to Old Work (illustrated by VW), where a focus on extrinsic compensation (e.g., salary, perks, status within a powerful company) and placing one's membership in an organization are at the center of one's life. Indeed, Frithjof Bergmann, often referred to as the founder of the New Work movement, described New Work as something more meaningful and beneficial for society in contrast to what he called "Job Work" (Bergmann, 2019), which we refer to as simply "Old Work." It is in essence, the idea that we could use technology "not to turn rivers into sewage or rain into acid, but for a wholly different (socially-beneficial) purpose" (Bergmann, 2019, p. 13). We could develop tools and processes that inject transparency into opaque processes and, in so-doing, help individuals find the best use of their talents for the larger benefit of society.

Of course, increasing transparency may require a major shift in corporate culture and society in general, which often prioritizes secrecy to cultivate competitive advantages (however short-lived). From this perspective, digital practice managers must be committed to bettering our society and lead with a strong moral compass rather than manage amorally. Amoral management refers to "failing to support a socially salient ethical agenda by not using ethical communication and not visibly demonstrating ethical practices" (see Greenbaum et al., 2015. p. 31). Managers must be clear that unethical forms of behavior, especially those that are pro-organizational (Mishra et al., 2021), are not to be tolerated.

A CASE STUDY OF OLD WORK CULTURE AND THE DARK SIDES OF IDENTITY AND COLLABORATION – THE VW EMISSION SCANDAL

The VW emission scandal is a major incident and a vivid example of corporate cheating and failed corporate governance. It inspired several case studies (e.g., Castille & Fultz, 2018; Jung & "Alison" Park, 2017; Mansouri,

2016; Prashant Singh, 2018; Wai et al., 2021). The various case studies of the VW emissions scandal generally agree that the root of this unethical scandal goes back to leadership, the longstanding organizational culture that was created and sustained over time, and structure of the company. Notably, this interesting case is an example of how managers and employees who identify strongly with their organization and may be considered good organizational citizens (i.e., going above and beyond for a company) in their context (see Murphy, 2021) work collaboratively to help their organization cheat to win.

Our chapter, which draws from work by Castille and Fultz (2018), summarizes the VW emissions scandal. This decade-spanning conspiracy was enacted by VW leadership with the goal of becoming the world's largest automaker. The conspiracy involved defrauding customers, investors, and government authorities in order to sell "clean diesel" vehicles to the U.S. and EU marketplace and involved the contribution of many employees both inside and outside of VW who over-identified with their organization. To avoid triggering the dark side of organizational identity and the collaborative cheating it may bring about, we encourage managers to adopt a values-based approach to implementing New Work philosophy.

Before we cover the VW emissions scandal in greater detail, it will be helpful to appreciate some key terms, particularly organizational identity and collaborative cheating. Identification is a process whereby an individual forms a perception of themselves as one with some larger collective (e.g., workgroup, organization, see Ashforth & Mael, 1989). Organizational identification, in particular, has a long history of academic study, with powerful effects on both positive outcomes (e.g., retention, job satisfaction, extra-role performance) and negative outcomes (e.g., unethical behaviors, resistance to change, poor performance, interpersonal conflict, negative emotions, and reduced well-being) (Conroy et al., 2017). Given this pattern of correlates, it is not surprising that organizational identification has been offered as a psychological explanation for the VW emissions scandal (Strauss, 2017, as cited by Murphy, 2021).

Additionally, collaborative cheating has been defined by Castille and Fultz (unpublished manuscript) as diverse actors coming together and interacting to create, implement, refine, and/or sustain dishonest or unfair activities intended to benefit a collective, the individual contributors, or both. Their definition combines accepted definitions of collaboration; that is, "the coming together of diverse interests and (or) people (e.g., varieties of expertise) to achieve a common purpose via interactions, information sharing, and coordination of activities," see Jassawalla & Sashittal, 1998, p. 239) and cheating (i.e., unethical acts, such as deception or trickery, that create unfair advantages for actors; see for example Mitchell et al., 2018). In other words, Castille and Fultz propose that collaborative cheating involves several individuals working together to gain an unfair advantage over rivals. This definition of collaborative cheating is analogous to corruption, which has been studied extensively by academics (e.g., normalization of corruption, Ashforth & Anand, 2003; Ashforth et al., 2008; moral disengagement as a cause of corruption, Moore, 2008; process models explaining how corruption emerges in organizations via social-psychological mechanisms, Palmer, 2008; Palmer & Maher, 2006; how emotions can cause corruption to spread like a virus, Smith-Crowe & Warren, 2014). However, their definition draws greater attention to the collaborative roots of corruption, which have not received as much attention until more recently (e.g., Weisel & Shalvi, 2015).

Research into collaborative cheating by Castille and Fultz (2018), upon which this chapter is largely based, illustrates how VW's collaborative cheating effort spanned over a decade and involved the contributions of actors (e.g., managers, executives, engineers, consultants) located throughout VW's network. We focus on key details to help our readers understand both the kind of toxic Old Work environment that organizational science suggests contributed to unethical behavior in the form of collaborative cheating. However, uniquely with this chapter, we view the scandal with a New Work lens, focusing on how the issue of over-identification (i.e., an individual's needs become almost entirely based on organizational membership; see Dukerich et al., 1998) that gave rise to the collaborative cheating behavior at VW might be avoided by practicing managers. This is important given the value that New Work places on identification with broader socially valued causes and collaboration (Bergmann, 2019).

Notably, Castille and Fultz (2018) conducted an in-depth qualitative research to create and then analyze a historical chain of key events that comprise the VW emission scandal. When Castille and Fultz (2018) were crafting the timeline of the VW emissions scandal, it quickly became apparent that the unethical behavior occurring at VW was not an isolated, one-time event (e.g., a single decision to defeat emissions testing). This is important because, very often, organizational leaders who are "caught up" in a scandal attribute bad behavior to a few isolated bad actors, which in fact was the case for VW. When the defeat device was discovered and VW executives were summoned to testify before the U.S. Congress, VW's then U.S. President and CEO Michael Horn suggested that a few rogue engineers installed the defeat device for "whatever reason" (Puzzanghera & Hirsch, 2015). Instead,

Castille and Fultz's analysis makes it clear that the collaborative cheating effort reflected many decision points faced by multiple VW employees from various departments and hierarchical levels, each of whom made unique contributions over a substantial period. It was a corrupt organization, not an organization that coincidentally had corrupt individuals within it (Pinto et al., 2008).

Castille and Fultz's (2018) qualitative approach highlights how a collaborative cheating effort can emerge over time, nurtured by organizational leaders willing to turn a blind eye to employee misdeeds so long as their highly aggressive goals (in this case, to be the world's largest automaker) are met. Castille and Fultz's approach appreciates the complexity of VW's collaborative cheating effort. They break down the history of the scandal into the following phases, each of which is defined by key events, some of which we summarize. For more details and a more thorough timeline, please see Castille and Fultz (unpublished manuscript). The first phase provides context for the creation (phase 1), implementation and refinement (phase 2), and failed concealment of the defeat device (phase 3).

CONTEXT – REGULATORY ENVIRONMENT SETS THE STAGE FOR A SORT OF ARMS RACE BETWEEN REGULATORS AND AUTOMAKERS, SUCH AS VW

1970s: Clean Air Act of 1970 (CAA) passed, establishing the existence of the Environmental Protection Agency (EPA) and equipping it with the power to penalize violators.

1973: VW is one of several automakers fined in the U.S. for using defeat devices to pass emissions testing.

1990s: CAA expanded its authority to enforce pollution regulations, which were also tightened. Regulations would undergo further tightening in the coming decades, placing even greater pressure on automakers to optimize between consumer demands for performance (e.g., power, fuel efficiency) and compliance (i.e., pollution).

1997: VW faces steep financial penalties for corporate espionage against its competitor, GM. VW Chairman and CEO Ferdinand Piëche had cultivated a reputation for ruthless leadership. In the industry, it takes no credit for wrongdoing.

Although the above event may appear unrelated to the most recent VW scandal, it provides important context for making sense of subsequent events occurring at VW. Notably, Piëche would later cede the CEO position prior to the scandal (2002) but remained as chairman till 2015, thereby maintaining some influence over VW leadership (notably, successor Martin Winterkorn).

PHASE 1 – THE DEFEAT DEVICE AT THE CENTER OF THE VW EMISSIONS SCANDAL IS CREATED/ASSEMBLED OVER TIME (*CREATION* OF THE DEFEAT DEVICE)

1999: Audi engineers create proto-software termed an "acoustic function" that is capable of defeating emissions testing. It is worth pointing out that Audi engineers were working on a project that is distinct from the project at the center of the VW emissions scandal.

2004: Robert Bosch (a consultancy) announced a push to convince U.S. automakers that diesel technology can comply with U.S. emissions standards, signaling their willingness to collaborate with VW to sell defeat device-equipped vehicles in the U.S. market.

2004–2007: After a series of failed efforts to optimize clean diesel technology for performance and regulatory compliance, VW engineers and managers, out of fear of losing their jobs or careers, collaborated across departments and with employees from Robert Bosch to adapt Audi's acoustic function to VW vehicles designed for the U.S. marketplace in order to defeat emissions testing. That the software in the car's electronic control module was designed to defeat emissions testing is beyond question. The defeat device worked by using data analyzing a series of factors, such as the speed, duration of engine operation, and barometric pressure to infer whether the vehicle was being driven on the road (and thus given the full power of the engine) or on a dynamometer (i.e., emissions testing conditions). The software by default assumed a vehicle was undergoing emissions testing at startup and switched to a "normal" setting if the barometric pressure rose to a level similar to that of driving on the road.

2005: VW, having violated the CAA once more, also becomes embroiled in a prostitution scandal.

The aforementioned details are important in this case as they provide evidence of a crucial collaboration between VW and Robert Bosch that is central to the emissions scandal as well as evidence of a culture that was permissive of questionable behavior. Although it should be noted that Bosch worked closely with VW to adapt the software to the engine destined for the U.S. market (the EA189 engine), a Bosch executive did write to VW, warning them that the use of a defeat device was prohibited by U.S. law and asked that VW indemnify Bosch against legal consequences, to which VW refused (Ewing, 2017). A manager from VW would go on to scold Bosch for getting lawyers involved (Ewing, 2017), and Bosch would go on to support VW's implementation and refinement of the defeat device in 2008 as issues began to arise.

PHASE 2 – VOLKSWAGEN MANAGEMENT APPROVES OF THE DEFEAT DEVICE (*IMPLEMENTATION* OF THE DEFEAT DEVICE)

2007: Martin Winterkorn becomes CEO of VW and issues Strategy 2018. Therein, Winterkorn establishes the most ambitious growth project for VW for a long time, calling on the VW Group (consisting of brands that include Audi and VW) to increase sales from six million per year to over ten million by 2018. In fact, Winterkorn's goals effectively required tripling U.S. sales by 2018. The strategy is to be executed by adding new models and, notably, expanding sales in the U.S., where VW historically performed rather poorly in selling diesel cars to consumers (diesel-engine vehicles made up just 5% of the U.S. car market at the time) (Brook et al. v. Volkswagen Group of America, Inc. et al., 2016). VW leadership, recognizing that the stigma surrounding diesel must be overcome (e.g., diesel engines emit thick, toxic smoke full of pollutants and created smog-filled cities in the past), go on to create a marketing plan to brand diesel as a clean, green alternative to hybrid engines, such as the Prius (Brook et al. v. Volkswagen Group of America, Inc. et al., 2016). For instance, VW created a website called "goodcleandieselfun. com" to change U.S. consumer perceptions about diesel.

2008: While the VW Engine Development Team touts the EA189 engine at a prestigious conference (29th International Vienna Motor Symposium), Winterkorn, at the same event, affirms VW's commitment to sustainability and innovation. Later in 2008 (May), a team leader, aware of the defeat device (engineer James Robert Liang), moved to the U.S. to assist with the launch of VW's new clean diesel vehicles. Liang, along with other co-conspirators, as part of the certification process, did not disclose the presence of the defeat device and deceived the EPA and CARB (California Air Resources Board). VW's clean diesel vehicles started selling in 2008. The use of the defeat device is referred to as an open secret at both VW and Bosch (Schmitt, 2016).

2012–2014: A series of hardware failures traced to the defeat device are identified by managers, prompting engineers to further modify the software and encourage managers to become complicit in the conspiracy to defeat emissions testing. Specifically, a steering wheel angle recognition is installed that determines whether the vehicle is being driven under regular operating conditions and shifts the vehicle out of "dyno mode" (Brook et al. v. Volkswagen Group of America, Inc. et al., 2016).

2014: German automakers financed a study in which VW took a lead role in polluting the science on the harmful effects of diesel emissions. Essentially, ten monkeys were forced to sit in airtight chambers and watch cartoons for entertainment as they were gassed by either a VW diesel-burning Beetle or a Ford diesel pickup. Although the study was never published, it reveals just how thoroughly corrupted VW was by this point in the scandal (Ewing, 2018).

PHASE 3 – VOLKSWAGEN'S COLLABORATIVE CHEATING EFFORT IS REVEALED DESPITE EFFORTS TO CONCEAL THE CONSPIRACY (*CONCEALMENT*)

2014–2016: Independent testing by the International Council on Clean Transportation reveals that VW's diesel-burning vehicles emit between 5 and 35 times the amount of toxins allowed by law. Regulators such as the EPA and CARB place increasing pressure on VW to provide answers for these discrepancies. Managers respond by restricting communications within VW and giving false reasons for compliance failures. In 2015, after a series of denials and phony recalls (where the defeat device was modified even further), a VW employee (Stuart Johnson) blew the whistle, contravening guidance from VW leadership and causing executives to admit the existence of the defeat device. Although indictments would follow in subsequent years, VW would nevertheless go on to become the world's largest automaker, overcoming Toyota in 2016 (Toyota would overtake VW five years later in 2021).

SUMMARY OF THE VW EMISSIONS SCANDAL

It is hard to overstate how VW's defeat device required individuals to collaborate to help their organization win in the marketplace by cheating. As others have argued (Murphy, 2021), it is quite clear that VW employees were over-identified with their organization and willing to do whatever it took to meet aggressive goals. VW leadership, most notably CEO Martin Winterkorn, established an aggressive goal focused on becoming the largest automaker on the planet without a clear strategy for realizing this goal. Winterkorn was a protégé of Ferdinand Piëche (a former VW CEO), a leader with a reputation for setting aggressive goals and firing engineers who could not find ways to meet them.

Such leadership behavior is a clear red flag for corruption in organizations (Taylor, 2016). For instance, according to one former VW worker, the software at the center of the scandal itself contained millions of lines of code (Milne, 2015), which is unlikely to have been produced by a single rogue engineer. Several reports make it clear that adapting this code to VW vehicles required employees from both VW and a software consultancy, Robert Bosch, to work collaboratively (Brook et al. v. Volkswagen Group of America, Inc. et al., 2016). Also, managers had several opportunities to raise concerns about the creation, implementation, and concealment of the defeat device (indeed, there are cases where managers did raise alarms) but chose to remain silent out of fear of reprisal (Ewing, 2017). In other words, VW was a thoroughly corrupt organization; it was a group that acted in a corrupt manner for the benefit of itself (Pinto et al., 2008). At least up until 2021, leaders and employees were still being prosecuted for their efforts to conspire and defraud customers, investors, and the public at large, some of whom serve prison sentences; meanwhile, VW has endured over $30 billion in fines and settlements (Ridley 2022). As a corrupt organization, it perhaps unsurprisingly spawned a collaborative cheating effort.

The excess pollution caused by VW's collaborative cheating effort is estimated to have non-trivial adverse impacts. These impacts include increased cases of chronic bronchitis (Barrett et al., 2015), excess hospital admissions, 120,000 days of restricted activity (e.g., lost work days), 210,000 days with lower-respiratory problems (U.S.), 33,000 days of increased asthma inhalers (U.S.), premature deaths in both the U.S. and Europe, and an ultimate economic impact of €1.9 billion, 60% of which occurs outside of Germany, and $450 million in the U.S. (see Barrett et al., 2015; Chossière et al., 2017).

Lessons for Managers Building a New Work Culture

The VW emissions scandal reveals the power of a corrupt organization to mobilize large swaths of workers, who likely identified strongly with their organization, to contribute to a collaborative cheating effort. Organizational identification is a potent force that can drive many positive outcomes. However, over-identification can be a cause of ethically questionable behavior and, when many others are involved in the same task, give rise to collaborative cheating. Such unethical behavior may be enacted either in the name of the company or merely to preserve one's own standing within the company.

A key takeaway for managers adopting a New Work philosophy is to build ethical infrastructures to ensure employees actively commit to doing the right thing, even if it harms the organization (Tenbrunsel et al., 2003). Ethical infrastructure is composed of both formal elements (e.g., communications that explicitly mention ethics, respect, and justice) and informal elements (e.g., strong climates that support ethical behavior, respect, and justice). Such infrastructure is unlikely to arise if managers adopt amoral stances (Greenbaum et al., 2015), which can be all too common among practicing managers.

One way to begin building this infrastructure is to adopt a values-based approach to decision-making whereby values are explicitly entered into the decision-making process. For instance, managers and their teams can undergo a values affirmation exercise where organizational values, which may include ethics, are sincerely affirmed either personally or within a group setting. Indeed, recent research suggests that by affirming deeply held values, the impact of performance pressure on cheating can be counteracted (Spoelma, 2021). Managers hoping to shift their culture from Old Work to New Work might start by finding out what parts of the Old Work culture will endure (e.g., certain core values), thus engendering a sense of continuity that can be helpful for overcoming resistance to change (see Conroy et al., 2017). Of course, de-emphasizing status differences by dismantling hierarchies and increasing transparency in decision-making can further encourage employees to shift into a New Work mindset.

Clearly, additional effort will be required to establish ethical infrastructures. Research has shown that an informal approach to ethical socialization was not as effective an impact to attitudes toward unethical behavior as a formal approach to ethical socialization (Mujtaba & Sims, 2006). This is not only a risk-management task but might imply a major change in the organizational structure and corporate culture. Less "job work" and more "New Work" might be necessary. The fast-changing working environment accelerated through digital transformation, and the 4th IR has tended to significantly increase the demands and pressure put on employees, often to the detriment of their health and personal life. Many organizations have expected more from their employees and have provided not much in return other than simply a job or to ensure employ-ability (Cartwright & Holmes, 2006). Employees, however, are increasingly looking for values and meaning in their jobs. According to a study from the Microsoft WorkLab, 53% of employees are more likely to prioritize health and well-being over work than before the COVID-19 pandemic (Microsoft, 2022). Hence, it might be an important action to revisit the company values and check that they are "up to the mark" and not only on paper but an inte-gral and vivid part of the corporate culture.

A key takeaway for policymakers and society could be that companies, for the sake of profit maximization, may not always follow very high ethical standards and may even break laws. An implication could be that rigorous enforcement is needed. U.S. law enforcement units initially triggered the observation that all investigations in this chapter mentioned that VW scandals might raise the concern if the European/German system is insuf-ficient to enforce regulations or if there are even (questionable) attempts to protect local companies. Loopholes in the regulatory framework and strap prosecution might contribute to repeated unethical actions in organizations and even provide fertile grounds for unethical behavior. Corporate social responsibility (CSR) has become an essential component of our current business world. Maybe the time is right to adapt the regulatory framework to support ethical standards of a changing business environment. It could help to prevent this kind of unethical behavior in the future and could even reduce costs for public health and society as a whole.

REFERENCES

Ashforth, B. E., & Anand, V. (2003). The normalization of corruption in organizations. *Research in Organizational Behavior, 25,* 1–52. https://doi.org/10.1016/S0191-3085(03)25001-2

Ashforth, B. E., Gioia, D. A., Robinson, S. L., & Treviño, L. K. (2008). Re-viewing organizational corruption. *Academy of Management Review, 33*(3), 670–684. https://doi.org/10.5465/amr.2008.32465714

Ashforth, B. E., & Mael, F. (1989). Social identity theory and the organization. *Academy of Management Review, 14*(1), 20–39.

Barrett, S. R. H., Speth, R. L., Eastham, S. D., Dedoussi, I. C., Ashok, A., Malina, R., & Keith, D. W. (2015). Impact of the Volkswagen emissions control defeat device on US public health. *Environmental Research Letters, 10*(11), 114005. https://doi.org/10.1088/1748-9326/10/11/114005

Bergmann, F. (2019). *New Work, New culture: Work We Want and a Culture that Strengthens us.* Zero Books.

Briggs, K., Workman, J. P., & York, A. S. (2013). Collaborating to cheat: A game theoretic exploration of academic dishonesty in teams. *Academy of Management Learning & Education, 12*(1), 4–17. https://doi.org/10.5465/amle.2011.0140

Brook et al. v. Volkswagen Group of America, Inc. et al. (United States District Court Northern District of California, San Francisco Division).

Cartwright, S., & Holmes, N. (2006). The meaning of work: The challenge of regaining employee engagement and reducing cynicism. *Human Resource Management Review, 16*(2), 199–208. https://doi.org/10.1016/j.hrmr.2006.03.012

Castille, C. M., & Fultz, A. (2018). How does collaborative cheating emerge? A case study of the Volkswagen emissions scandal. Hawaii International Conference on System Sciences. https://doi.org/10.24251/HICSS.2018.014

Castille, C. M., & Fultz, A. (unpublished manuscript). Elaborating on the collaborative roots of corruption: Insights from the Volkswagen emissions scandal.

Chossière, G., Malina, R., Ashok, A., Dedoussi, I., Eastham, S. D., Speth, R., & Barrett, S. R. H. (2017). Public health impacts of excess NOx emissions from Volkswagen diesel passenger vehicles in Germany. *Environmental Research Letters.*

Conroy, S., Henle, C. A., Shore, L., & Stelman, S. (2017). Where there is light, there is dark: A review of the detrimental outcomes of high organizational identification. *Journal of Organizational. Behavior, 38,* 184–203. doi: 10.1002/job.2164.

Dukerich, J. M., Kramer, R., & Parks, J. M. (1998). The dark side of organizational identification. *Identity in Organizations: Building Theory Through Conversations,* 245–256.

Ewing, J. (2017). *Faster, Higher, Farther: The Volkswagen Scandal.* W. W. Norton & Company.

Ewing, J. (2018, January 25). 10 monkeys and a Beetle: Inside VW's campaign for "Clean Diesel." *The New York Times.* www.nytimes.com/2018/01/25/world/europe/volkswagen-diesel-emissions-monkeys.html

Greenbaum, R. L., Quade, M. J., & Bonner, J. (2015). Why do leaders practice amoral management? A conceptual investigation of the impediments to ethical leadership. *Organizational Psychology Review, 5*(1), 26–49. https://doi.org/10.1177/2041386614533587

Jassawalla, A. R., & Sashittal, H. C. (1998). An examination of collaboration in high-technology new product development processes. *Journal of Product Innovation Management, 15*(3), 237–254. https://doi.org/10.1016/S0737-6782(97)00080-5

Jung, J. C., & "Alison" Park, S. B. (2017). Case study: Volkswagen's diesel emissions scandal. *Thunderbird International Business Review, 59*(1), 127–137. https://doi.org/10.1002/tie.21876

Mansouri, N. (2016). A case study of Volkswagen unethical practice in diesel emission test. *International Journal of Science and Engineering Applications, 5*(4), 211–216. https://doi.org/10.7753/IJSEA0504.1004

Martin, S. R., Kish-Gephart, J. J., & Detert, J. R. (2014). Blind forces: Ethical infrastructures and moral disengagement in organizations. *Organizational Psychology Review, 4*(4), 295–325. https://doi.org/10.1177/2041386613518576

Microsoft. (2022). *2022 Work Trend Index: Annual Report.* https://ms-worklab.azureedge.net/files/reports/2022/pdf/2022_Work_Trend_Index_Annual_Report.pdf

Milne, R. (2015, October 6). *Culture clash gives clue to Volkswagen scandal. Financial Times.* www.ft.com/content/24a9c128-6b5a-11e5-8608-a0853fb4e1fe

Mishra, M., Ghosh, K., & Sharma, D. (2021). Unethical pro-organizational behavior: A systematic review and future research agenda. *Journal of Business Ethics.* https://doi.org/10.1007/s10551-021-04764-w

Mitchell, M. S., Baer, M. D., Ambrose, M. L., Folger, R., & Palmer, N. F. (2018). Cheating under pressure: A self-protection model of workplace cheating behavior. *Journal of Applied Psychology, 103*(1), 54–73. https://doi.org/10.1037/apl0000254

Moore, C. (2008). Moral disengagement in processes of organizational corruption. *Journal of Business Ethics, 80*(1), 129–139. https://doi.org/10.1007/s10551-007-9447-8

Mujtaba, B. G., & Sims, R. L. (2006). Socializing retail employees in ethical values: The effectiveness of the formal versus informal methods. *Journal of Business and Psychology, 21*(2), 261–272. https://doi.org/10.1007/s10869-006-9028-3

Murphy, K. R. (2021). *How Groups Encourage Misbehavior.* Routledge.

Palmer, D. (2008). Extending the process model of collective corruption. *Research in Organizational Behavior, 28*, 107–135.

Palmer, D., & Maher, M. W. (2006). *Developing the Process Model of Collective Corruption.* Sage Publications Sage CA: Thousand Oaks, CA.

Pinto, J., Leana, C. R., & Pil, F. K. (2008). Corrupt organizations or organizations of corrupt individuals? Two types of organization-level corruption. *Academy of Management Review, 33*(3), 685–709. https://doi.org/10.5465/amr.2008.32465726

Prashant Singh, P. S. (2018). Volkswagen emissions scandal—A case study report. *International Journal of Human Resource Management and Research, 8*(5), 11–18. https://doi.org/10.24247/ijhrmroct20182

PricewaterhouseCoopers. (2020, October 29). Digital ethics necessary for successful digital transformation. PwC. www.pwc.nl/en/topics/blogs/digital-ethics-necessary-for-successful-digital-transformation.html

Puzzanghera, J., & Hirsch, J. (2015, October 8). VW exec blames "a couple of" rogue engineers for emissions scandal. www.latimes.com/business/autos/la-fi-hy-vw-hearing-20151009-story.html

Ridley, K., (2022, May 25). Volkswagen in $242 mln UK "Dieselgate" settlement. *Reuters.* www.reuters.com/business/autos-transportation/volkswagen-242-mln-uk-dieselgate-settlement-2022-05-25/

Schmitt, B. (2016, January 23). Dieselgate crown witness: cheating open secret At VW. Forbes. www.forbes.com/sites/bertelschmitt/2016/01/23/dieselgate-crown-witness-cheating-open-secret-at-vw/#37a96d0e5002

Smith-Crowe, K., & Warren, D. E. (2014). The emotion-evoked collective corruption model: The role of emotion in the spread of corruption within organizations. *Organization Science, 25*(4), 1154–1171. https://doi.org/10.1287/orsc.2014.0896

Spoelma, T. M. (2021). Counteracting the effects of performance pressure on cheating: A self-affirmation approach. *Journal of Applied Psychology*. https://doi.org/10.1037/apl 0000986

Strauss, K. (2017, July 26). How Volkswagen rallied its employees after its emissions scandal (At least for now). *Forbes*. www.forbes.com/sites/karstenstrauss/2017/07/26/how-vol kswagen-rallied-its-employees-after-its-emissions-scandal-at-least-for-now/

Taylor, A. (2016). *What Do Corrupt Firms Have in Common? Red Flags of Corruption in Organizational Culture*. Center for Advancement of Public Integrity, Columbia Law School. www.law.columbia.edu/CAPI

Tenbrunsel, A. E., Smith-Crowe, K., & Umphress, E. E. (2003). Building houses on rocks: The role of the ethical infrastructure in organizations. *Social Justice Research*, *16*(3), 285–307. https://doi.org/10.1023/A:1025992813613

Wai, C. K., Teck, T. S., Junkai, M., Lu, H., Hoo, W. C., Hong Ng, A. H., & Sam, T. H. (2021). Mapping the Volkswagen diesel dupe crisis, its implications and sustainability of their responses. *International Journal of Academic Research in Business and Social Sciences*, *11*(5), Pages 211-226. https://doi.org/10.6007/IJARBSS/v11-i5/9919

Weisel, O., & Shalvi, S. (2015). The collaborative roots of corruption. *Proceedings of the National Academy of Sciences*, *112*(34), 10651–10656. https://doi.org/10.1073/pnas.142 3035112

Welsh, D. T., Ordóñez, L. D., Snyder, D. G., & Christian, M. S. (2015). The slippery slope: How small ethical transgressions pave the way for larger future transgressions. *Journal of Applied Psychology*, *100*(1), 114–127. https://doi.org/10.1037/a0036950

6

How Industry 4.0 Influences Our Work Environment

David Galipeau
SDGx Pte. Ltd, Singapore

CONTENTS

SETTING THE STAGE

Innovation drives economic growth for all organizations—multinationals and medium-sized enterprises alike. It is also a necessary focus of policy-making and civic engagement in an era of rapid change due to the new and emerging Industry 4.0 technologies, which in this chapter I will refer to as IR4.0. There are varied definitions for IR4.0, but I prefer the *Fourth Industrial Revolution, 4IR, or Industry 4.0, which conceptualizes rapid change to technology, industries, employment, and societal patterns and processes due to increasing interconnectivity and automation of traditional manufacturing and industrial practices* (Bai et al., 2020). It blurs boundaries between the physical, digital, and biological worlds because of the convergence of several new

DOI: 10.1201/9781003371397-6

FIGURE 6.1
A robot evaluating a CV or interviewing a human.

and innovative technologies. Each industrial revolution, apart from techno-logical changes, has resulted in both positive and negative economic and social disruptions.

This is true today. Businesses everywhere face increased pressure to trans-form the current consumer demand-centric business models into new innovative business models, often taking on additional risks to adapt to a new digitalized economy, fast-changing consumer behaviors, and creating new markets with unknown value. As a result, innovating has become increasingly complex. This complexity is compounded as increased collab-oration between non-traditional partners and industry ecosystems emerges. Innovation previously had the competitive advantage of well-financed research and development teams. Innovation in the near future will increas-ingly rely on collaborative platforms that service individual entrepreneurs, new startups, and traditional small and medium-sized enterprises (SMEs)—the backbone of many nations' economic growth and one of the largest employers globally (Clark, 2021).

The 2008 financial and economic crisis illustrated that SMEs prove to be more robust than large and multinational enterprises due to their agility, entrepreneurial flexibility, and competitive innovation capabilities. Integrating

IR4.0 technologies into product manufacturing and service operations will lead to SME 4.0—SMEs that adopt IR4.0 technologies, disrupting supply chains and operational relationships. In doing so, foundational challenges still exist; for example, internet access in remote areas, cybersecurity, e-waste, appropriately skilled labor markets, startup competition, and even environmental, social, and governance (ESG) regulations. The SME value proposition is highly skilled workers—human capital—but, as most of us may be experiencing, the innovation strategies for many organizations—private, public, and civic —focus primarily on adopting IR4.0 technologies. Additionally, the fundamental reason for leveraging IR4.0 technologies into actual practice is to manage risk (business continuity in disruptive environments), increase competitiveness (market expansion and creation), and operational efficiencies (value creation). The private sector needs to remain competitive to survive. Still, the public sector (nation-states) must also compete regionally and globally for trade opportunities, foreign direct investment, foreign talent, and technology transfers to increase domestic resilience and reduce the risk of socioeconomic disruptions at home.

Striking a balance between fair and decent working conditions and unlocking the innovative potential of IR4.0 is a crucial challenge for policymakers. IR4.0 technologies are revolutionizing the manufacturing industry, enabling faster production times at lower costs with less human interaction, but IR4.0 also entails the infusion of higher-value-added processes through digitization, advanced technologies, and efficient resource utilization as outlined by various sustainability protocols such as ESG[1] principles and the United Nations Sustainable Development Goals (UN SDGs)[2]. Yet, on many occasions, large-scale implementations of technological solutions do not deliver the "planned" positive impacts in productivity, cost, quality, or reliability.

Although media and most academic research focus on how IR4.0 technology can help all businesses to remain competitive, there is little exploration of the impact on other stakeholders and, more to our point, the future implications for work, workers, and the workplace. For example, according to the Boston Consulting Group forecast, the rate of automation of manual tasks will increase exponentially worldwide, reaching almost full saturation by the end of the 2020s (Zinser et al., 2015). In my experience, working to advance IR4.0 technologies in both the private and public sectors of developed and developing nations, the challenge is not finding the practical applications or the investment capital for IR4.0 technology but a lack of accompanying investments in the most critical asset: people (Carpi et al., 2019).

By exposing this investment gap, we can understand why there may be misunderstandings about the full impact of IR4.0 technologies on society, the environment and, perhaps most importantly, human capital. Understandably, the situation is now very different—the workers now are not the same as those before the pandemic. The pandemic experience of the past two years left an indelible mark on the psyche of workers, altering their expectations and fundamentally shifting the way work gets done.

This is a critical moment where leadership matters more than ever. Those who embrace a new mindset and shift cultural norms will best position their people and business for long-term success.

Adapting to these new expectations is good for workers. Still, it can also be a competitive advantage that ultimately impacts business continuity, allowing HR departments to upskill current employees and attract new talent to their ranks.

While the labor market transformation has been underway for decades, rapid technological advancements and, more recently, the COVID-19 pandemic have expedited the trend and compelled policymakers to devise adequate responses.

CASE: SINGAPORE SMART INDUSTRY READINESS INDEX

Singapore is a recent example of how a nation can implement IR4.0 technologies with holistic impact. Since its independence in 1965, Singapore has grown wealthy by opening its borders and encouraging financial institutions to locate there. However, the Asian financial crisis (1997–08), which saw Singapore experience its first significant economic downturn in decades, led to a change in public policy strategy toward achieving economic diversification for future risk mitigation. Economic diversification remains a crucial challenge for most countries. Still, in the case of Singapore, there were many benefits, including jump-starting its national digital transformation journey and rebalancing the relationship between the public, private, and civic sectors.

On November 13, 2017, the Singapore Economic Development Board (EDB) launched the Smart Industry Readiness Index (SIRI), which promoted a suite of framework tools to help manufacturers start, scale, and sustain their digital transformation journeys. SIRI covers the three core elements of Industry 4.0: Process, technology, and organization, and over 350

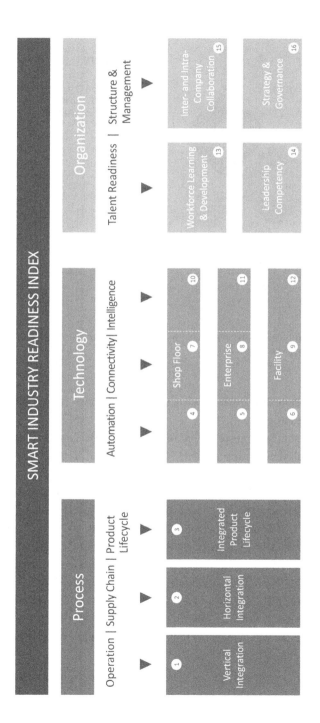

FIGURE 6.2
SIRI matrix. SIRI identifies three fundamental building blocks of Industry 4.0: process, technology, and organization. (Adapted from Singapore Economic Development Board [Singapore EDB], 2019b).

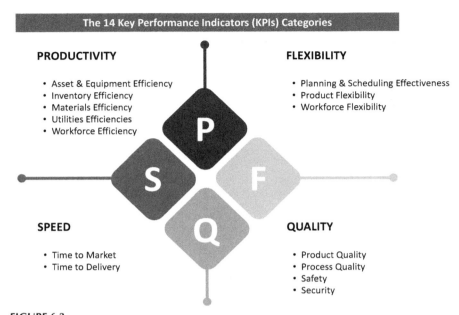

FIGURE 6.3
The 14 key performance indicator categories (adapted from Singapore Economic Development Board [Singapore EDB), 2019a].

Singapore-based companies from more than 15 countries have completed the Official SIRI Assessments.

In October 2019, the Singapore EDB announced a partnership with the WEF to drive the global adoption of SIRI. The EDB and the WEF will accelerate the use of SIRI as an internationally recognized standard for Industry 4.0 transformation and the adoption of new IR4.0 technologies across manufacturing communities worldwide. It has since been adopted internationally by SMEs and large multinational corporations.

To develop a holistic IR4.0 transformation strategy and roadmap, firms must consider all three building blocks of the SIRI framework (process, technology, and organization). The prioritization matrix considers the 14 KPI categories (Singapore EDB, 2019a).

Three KPIs specifically outline monitoring and evaluation criteria for measuring the worker impact of IR4.0 technology implementation projects. These are:

KPI 2: Workforce Efficiency	KPI 8: Safety	KPI 12: Workforce Flexibility
KPIs under this category assess both the direct and indirect labor productivity in factories or plants. Companies with solid workforce efficiency will generally require fewer person-hours per task. Depending on the companies' HR policies, employee turnover and training effectiveness could also be additional parameters in both influencing and determining workforce efficiency. Improvements in workforce efficiency can result in more remarkable revenue contributions per employee and enhance intangible elements like workforce morale.	KPIs under this category measure and track the number of workplace health and safety incidents in the company's facilities. Having few or no health and safety issues within the working environment will help minimize disruptions and ensure the company's operations continue running smoothly. Additionally, keeping employees healthy and safe reduces regulatory and compensation costs and, more importantly, helps raise the morale and confidence of employees. This enabling space can also create positive spin-offs in employer branding and customer confidence in the long run.	KPIs under this category assess the ability of a company's workforce to perform a variety of different job functions. Building a solid team of multi-skilled workers across the different levels of the organization allows a manufacturer more options in potentially redeploying employees during periods of volatility, such as when market demand fluctuates. This adaptability enables the company to achieve stability more quickly and effectively. Having a flexible workforce also allows a company to cope better with sudden attrition, thus strengthening the company's business continuity.

IS THIS ENOUGH?

From the mechanical loom to the full-scale automation, no technology can achieve anything by itself. Technology can act "smart," but it does not have the human characteristics of intelligence, wisdom, patience, or reflection. Technology itself does not know if it is acting morally, ethically, inclusively, or equitably. This is important because change is happening faster than expected.

Over the past decade, the global sales of industrial automation platforms tripled, with annual growth of 58% during the forecast period from 2020 to 2027 (Coherent Market Insights, 2021).

The immense growth in industrial robots is exceptionally high in countries with vital automotive and electronics sectors, namely Japan, China, South Korea, Germany, and the United States. In South Korea, there are over 930 robots per 10,000 manufacturing employees—seven times more than the world average.

The SIRI framework recognizes that more value is created by integrating both the technologies and the people who manage them. Still, as the digital transformation accelerates, new models of capability building are essential, requiring even further investment in human capital—from senior managers to factory floor workers. All staff will need to become comfortable with complex and sophisticated new technologies and mold them into tools they can leverage. New and innovative solutions often fail to produce convincing results because the workforce does not understand or trust them. Therefore, senior managers developing and implementing IR4.0 strategies should start asking questions about their investment mix and how they prioritize human capital investment for the years ahead. The importance of human capital investments is not decreasing in IR4.0—I believe the opposite is true. Rising productivity will result in smaller workforces so that each remaining employee will have a more significant impact on aggregate performance. It will increase the demand to develop skills, capabilities, and, perhaps more importantly, mindsets.

We can assume that many current jobs will be automated, but that process will take decades. Companies exploring automation will still rely heavily on their workforce until at least the middle of the century and, after that, they will still need human capital to manage, adapt, and optimize their automated assets. Humans will remain at the core of productivity increases in most industries for the foreseeable future. Recent research published by the School of Management del Politecnico di Milano, the Osservatorio Industria 4.0 suggests that, to maintain the centrality of the human factor within the new contexts, IR4.0 companies must increase their skills in HR selection, training, and leadership development (Osservatorio Transizione Industria 4.0, 2020).

Therefore, Skills 4.0, or a broader Education 4.0, becomes a critical success criterion of the era of IR4.0. Organizations need to reflect on how significant the new technology trends will be in maintaining staff training programs to remain competitive. As in past technology evolutions, a successful IR4.0 transformation depended on the willingness of senior management to create the enabling space for change and the capacity for employees to embrace it. As the workforce upgrades its skills and transforms, it will discover new ways to connect and share knowledge, co-create ideas, identify new data sources, and apply tools in creative ways.

This increases workforce value as organizations shift from implementing disconnected IR4.0 technologies across many departments to a cohesive model leveraging IoT and ultimately to the internet of systems (IoS)—recently referred to as the system of systems. This organization-wide willingness to accept and adapt to change is imperative for the successful application of IR4.0 technologies, and this is the nature of human-centric systems thinking.

THE SHIFT FROM THE INTERNET OF THINGS TO THE INTERNET OF SYSTEMS – A HUMAN-CENTRIC APPROACH

The basics of the IoT describe physical objects (or groups of such objects) embedded with sensors, processing ability, software, and other technologies that connect and exchange data with other devices and systems over the internet (Gillis, 2022), but the term "Internet of Things" is a misnomer because devices do not need to connect to the public internet—they only need to be connected to a computer network and be individually addressable (Middleton et al., 2013).

As technologies such as ubiquitous computing, sensors, embedded systems, and machine learning interoperate to create an IoT platform, they become synonymous with "smart" products. These can include lighting, heating, cooling, and security products that support smart appliance ecosystems. Users control these smart appliances via voice, gesture activation, or automated AI devices, such as smartphones and smart speakers.

IoT is also the underlying technology (together with 5G) for smart cities, smart cars, smart factories, smart infrastructure, and smart healthcare systems (Laplante et al., 2018). But as IoT applications expand from a network of interconnected "things" to a network of connected systems, several privacy and cybersecurity risks become evident. Industries and governments address these risks by developing international standards, guidelines, and regulatory frameworks. Unfortunately, much of the focus is on citizens' privacy or consumer protection—workers are often under-prioritized. Even when describing risks such as privacy, transparency, and trust, often these descriptions remain technical, seeing them as technological and not social challenges. For example, the internet was initially described as networking technology but it is now evident that the internet is essential for delivering many vital knowledge and welfare services such as online education platforms, social media communications, and healthcare. Many work remotely using video communications and other online tools in the post-pandemic era. The data gaps identified during the pandemic reveal the biased relationship between those who collect, store, and analyze data and those the data is meant to benefit. At the extreme, this exposes a new type of data class—the data-poor— those with little to no access to online networks. Marginalized communities include undocumented migrants, refugees, and undocumented labor operating under insecure conditions, such as sex workers, gig workers, and farmhands.

It is clear to most that IoT will be at the center of the digital revolution. Still, by including the impact on workers and citizens, there is an opportunity to develop systems that prioritize human-centered IoT solutions. A systems approach can create meaningful change but we need to identify system-level challenges that impact all stakeholders—users, citizens, workers, consumers, and voters—and motivate them to collaborate and co-create long-term solutions.

As Artemis Research suggests, a human-centric IoT strategy may include:

- Start the IR4.0 transformation with solid leadership support of a human-centric vision.
- Use of IoT as a means for creating and maintaining valuable human-machine relationships.
- Focus on improving human-machine interfaces and usability.
- Leverage IoT to complement and improve human efficiency, reduce workload and simplify complex tasks (Gardoni, 2020).

Human-centered system design is also valuable to support policymakers that focus on the changing nature of work and labor markets and how best to prepare people and societies for future jobs. In the years ahead, IR4.0 technologies—specifically IoT-related automation—will have a substantial economic and social impact on nations worldwide, and governments must play an important role. COVID-19 brought massive disruption to the workforce, stimulating new business model development and consumer behavior changes, many of which are likely to endure. The accelerated post-pandemic trend in remote work, education, health, e-commerce, and automation could result in up to 25% more workers than previously estimated needing to switch occupations (Lund et al., 2021). In systems that operate, maintain, or communicate with people, the human factor is critical and has unique capabilities and limitations (Watson et al., 2020).

Human intelligence is a critical factor for IoT, enabling the shift from IoT to IoS and upgrading the platforms to integrate seamlessly with human activities on the work floor and off. Human intelligence allows IoT to evolve from simple data collection to knowledge collection—a value proposition for systems engineering that connects four categories of users:

1. Inexperienced users (a person who uses healthcare monitoring devices at home).
2. Experienced users (a maintenance operator that repairs IoT-networked industrial machinery).
3. Superusers (the administrator or operator managing the IoT infrastructure).
4. The designers and developers of IoT interfaces, applications, and user experience.

Here we see that a human-centric systems approach looks at the complete workflow from developer to user with feedback loops to maintain agile stability.

Another example of the IoT human-centric approach leveraging virtual or AR platforms for knowledge sharing may include:

- Education and professional training based on virtual, AR, and total immersion spatial computing applications, including the metaverse and multiverse.
- Augmented site inspection (i.e., safely guide operators to faulty equipment).

- Simplified maintenance operations (i.e., extended reality hardware facilitates hands-free operation, and the IoT connections during repairs provide access to reference sources, like live diagnostic feeds, technical manuals, coworkers, and help desks).
- Human-assisted robotic manufacturing (i.e., AR allows the operator to select the parts to pick up and assemble, then IoT takes over and starts the complete robotics-based manufacturing process).
- Simplified fleet management (i.e., augmented 3D-based asset management, windshields that provide traffic data, weather alerts, cargo status).
- Improved worker safety and efficiency (i.e., safety helmets that provide task lists, safety instructions, surveillance, monitoring vital signals).
- Remote support for first responders.

SHIFTING FROM SMART TO INTELLIGENT

When discussing IR4.0, we tend to focus on technology or digital tools to solve challenges. But how can organizations move beyond technology focus alone and shift from smart to intelligent solutions?

As we previously stated, human intelligence allows IoT to evolve from simple data collection to a more valuable knowledge collection. This insinuates that human knowledge becomes the essential requirement to shift from smart "things" to intelligent "systems", enabling a complete transition from the information society to the knowledge society or, perhaps, the intellectual society.

As IR4.0 technologies like AI/machine learning, robotic process automation, and cognitive computing threaten to disrupt business models, HR management must adapt to a hybrid model of labor that includes a level of diversity not seen previously. Leadership will need to shift from the traditional "command and control" function to a new level of empowering and enabling staff.

Many jobs will be redefined, redesigned, recombined, or simply disappear. Many workers will lose their jobs, and new jobs will be created. However, the time lag between these two states will increase due to the advanced sophistication of IR4.0 technologies. In contrast to the internet or dot-com boom of the late 90s, IR4.0 technologies cannot be taught in a few weeks, so the latency between job losses and acquiring the necessary skills for new IR4.0 jobs may take a long time, if achieved at all.

Jobs affected will be blue-collar and white-collar professions such as medical, legal, and finance professionals. However, there is a silver lining. Automation will replace mundane and routine tasks that may open opportunities for displaced workers with updated skills to undertake higher-value work and potentially offset the predicted loss of middle-income jobs (Neal & Caplan, 2018).

The speed of job losses will not be as fast as previously expected because current automation platforms are not agile and intelligent enough to replicate complex tasks such as making a "judgment call"—a decision based on reflection and tacit knowledge that cannot be easily codified.

As a requirement to shift from smart to intelligent, organizations will need to fully understand how IR4.0 technologies can cohabitate with human workers to uncodified experience and knowledge. This will be incredibly challenging for HR departments that may have to recruit human and non-human talent in the future.

SIRI illustrates that industry readiness is critical, but worker readiness may be more vital to students, workers, researchers, educators, organizational trainers, HR professionals, and policymakers. Recent research (Blayone & VanOostveen, 2021) has summarized five activity-system characteristics of worker-level readiness factors.

Technological Readiness	Basic digital skills develop through regular use of mainstream hardware (personal computers and mobile devices) and software/apps. Advanced levels include: • networking and information processing • data analysis and working with raw materials, smart objects, automated guided vehicles (AGVs), and machinery • ability to work with complex software interfaces where additional interpersonal skills such as high technology acceptance and low technology anxiety are required
Interpersonal Readiness	Focus on cognitive, processual competencies and soft skills such as: • social negotiation skills to achieve win-win outcomes • cross-cultural and online communication competencies for working effectively within geographically dispersed teams • capacity to respond situationally with affective computing and social robots

Flexibility Readiness	Adaptability is shaped by cultural and personal dispositions such as:

- multidisciplinary knowledge
- cognitive flexibility and environmental adaptiveness
- openness to dynamic roles and emergent problems
- comfort with technological change
- tolerance of ambiguity

Inter-agent Readiness	Interacting with machines happens mainly in two ways: Tasks performed by human operators with machine supervision or tasks performed by machines with human supervision. These interactions differ between agents, including human, non-human, and hybrid entities. Some worker competencies may include:

- ability to achieve optimal levels of comfort and performance within tightly integrated human-machine assemblages
- openness to human-machine partnering
- trust toward technological entities, including robots, decision-automation systems, big-data analytics, and augmentation apparatuses
- ability to model the functioning of non-human agents
- communicate in ways that non-human agents can readily process

Innovation Readiness	Although creative problem-solving is expected to evolve as a hybrid process incorporating humans and automation systems, innovative thinking and creative intelligence are still challenging areas for machine automation; therefore, worker competencies may include:

- establishing an enabling space for experimentation
- generating practical ideas
- refining and evaluating varied combinations of ideas
- navigate unexpected or irrational challenges
- functioning collaboratively
- taking risks, making mistakes, and learning from failures

Considering this workers' readiness index, we can suggest critical shifts in strategic human-centric approaches:

- Systematically scrutinizing current operational processes to identify the risks of automating labor and the implications on recruitment, training, and capacity-building.
- Introducing system-wide digital literacy training programs to build the capacity required to advance competitiveness, identify new opportunities, mitigate challenges, and reduce job losses.
- Enable a culture for self-directed learning while introducing support programs for existing and new staff and management.
- Design and implement leadership development programs focusing on hybrid team dynamics.

Historically, HR policies and leadership support are designed to support on-site workers, but new hybrid models are an opportunity to accelerate IR4.0 technologies. A hybrid workforce can work from the company's designated office and remotely. The workforce can be a human, a machine, or a mixture of both. Working within a hybrid environment has become more common since the pandemic, as organizations seek to remain competitive by offering staff more flexible WFH options.

As Gartner, a technology research and consulting firm based in the USA recently published:

> COVID-19 has shattered the paradigm of in-office work. Once a consideration for many organizations but rarely a priority, remote work has become a health and safety imperative. Even as organizations plan their recovery strategies, however, remote work will remain a cornerstone of the post-pandemic future of work.
>
> Gartner 2022

Shifting from smart to intelligent is not only about smart cities, smart cars, and smart factories; it is how organizations introduce IR4.0 technologies that complement and raise the effectiveness and efficiencies of current staff and workers.

IR4.0 technologies have become enablers for a new social contract between employers and employees, resulting in a more equitable and inclusive working environment with a high potential for both social and financial returns.

SOCIETY 5.0

An interesting case is Society 5.0, a selection of public policies introduced in 2020 by Japan's Council for Science, Technology, and Innovation. Japan seeks to cultivate a Society 5.0 that incorporates IR4.0 technologies in all industries and social activities to achieve economic growth and solutions to social problems in parallel. In Japan's public sector, a priority has previously been placed on economic systems. As the digital transformation quickly advances, cracks in worker, consumer, and citizen contentment and well-being have been identified. Society 5.0 takes a whole-of-society or systems approach of the digital transformation journey to counter this. It attempts to achieve a balance between cyberspace and physical space to support all citizens. For the evolving digital labor force in Japan, this eventually will free workers from work and tasks that they are not particularly good at, thereby optimizing workers' social and organizational systems (Cabinet Office, n.d.).

For example, let us look at how the Society 5.0 program seeks to create new value chains in Japan's manufacturing sector by introducing AI analysis of big data across diverse types of information. By doing so, the following is expected to be realized:

- Perform flexible production planning and inventory management in response to current needs by establishing links with other fields and industry suppliers outside of one's usual business dealings.
- Use AI and robots and apply inter-plant coordination to make production more efficient, save on labor, enable the inheritance of technical skills (master craftsmanship modeling), and achieve high-mix, low-volume production.
- Make distribution more efficient by cross-industry cooperative shipping and truck platooning.
- Enable customers and consumers too to obtain low-priced goods without delivery delays.

Furthermore, for society overall, these solutions can help strengthen industrial competitiveness, enhance responsiveness to disasters, mitigate the labor-shortage problem, deal with diverse needs, reduce greenhouse gas emissions and expenses, improve customer satisfaction, and create new upskilling opportuning and jobs.

The McKinsey Global Institute estimates that 40–50% of Japan's full-time equivalents will be replaced by automation between 2016–2030 (Manyika et al., 2017). Therefore, job creation and upskilling are essential in Japan.

To counter this potential disruption, Japan's Ministry of Education, Culture, Sports, Science, and Technology (MEXT) has created the Human Resource Development for Society Task Force on Developing Skills to Live Prosperously in the New Age to explore new capacity-building principles. These seek to "transcend the humanities/sciences divide" (MEXT, 2018) to create a workforce that can complement the new IR4.0 technologies. These efforts also include the civil service workers within the public sector as it transforms the traditional ministerial bureaucracies by leveraging IR4.0 technologies as valuable tools for analyzing the process of policymaking in complex environments and the delivery of inclusive and equitable digital public services. Although we have seen that the SIRI approach manages workforce challenges, Japan's Society 5.0 moves toward a whole-of-society system. In this process, finding talented and qualified employees and ensuring their commitment, availability of training and career development opportunities, and leaders' ability to coach their employees are considered essential points. New science and technology departments are being created at universities to develop new talents to support the Society 5.0 transformation to ensure that workers are fit for purpose.

A vital success criterion of Society 5.0 is the existence and availability of talents, career planning, talent management, comprehensive leadership development, and the diversity of workers; therefore, HR departments should include Society 5.0 in their HR development and management strategies.

By taking advantage of unique factors, including Japan's advanced technology cultivated from "*monozukuri*", or Japan's excellence in the manufacturing of things, and years of basic research, Japan seeks to overcome social challenges such as a decrease in the productive-age population, aging of local communities, and energy and environmental issues ahead of other nations.

Japan believes that Society 5.0 will create an abundance of new employment opportunities by creating a vibrant economic society, improving productivity, and creating new markets. By doing this, Japan will play a key role in expanding the new Society 5.0 model to the world.

Society 5.0 places greater value on human intelligence and seeks to foster a more balanced working relationship between IR4.0 technologies and humans. Workers do not compete with robots for jobs in this scenario but rather collaborate. To realize this, workers will need to engage in the discussions collectively.

IS IT TIME FOR DIGITAL UNIONS?

In the post-pandemic era, workers are more particular about what they will and will not tolerate at work, about what they want their business to stand for, and are more confident in expressing their wishes. Although workers' opinions are being considered at times, they need to engage within the process and make their concerns known, but they cannot do this alone.

Let us explore alternatives to the formal or informal, top-down approach to behavior change—how IR4.0 technologies can rebuild the traditional workers' support organizations; for example, labor unions. Historically, workers and their work were tied to fixed locations—an office, warehouse, factory, city, country—but with the increase in internet penetration, even in the most remote and challenging environments, workers and their work have been decentralized. The result is that the new world of digital work, enabled by IR4.0 technologies, is restructuring power relations between capital, policy, and labor. Labor unions representing workers in the digital transformation—and that adopt IR4.0 technologies—can help to facilitate multi-stakeholder dialogues to address access to social protection, digital and informal economy worker representation, algorithmic management, and transparency and equitable access to upskilling and re-skilling.

Traditionally, a labor union enables workers to exercise the right of collective bargaining to negotiate wages, benefits, and working conditions. As an example, to create a collective movement for job security, union pay, and better working conditions, Amazon workers have created the first completely independent worker-led union in decades, if not centuries. The movement leader, and former Amazon worker Mr. Chris Smalls, who made his name protesting for safety conditions at the retail giant during the pandemic stated, "We did whatever it took to connect with these workers and relied on an online fundraiser."[3]

This may be the next evolution of the "working class" for both the white-collar professions and the blue-collar tradespeople and a rebalancing of the employee-employer social contract.

In the post-pandemic era, labor unions are in disruption mode globally. Manufacturers are driving more automation based on increasing efficiency and competitive risk rationalization. Service sectors are being transformed using mobile apps, social media platforms, data analytics, and the increasing application of AI. Cross-border digital companies have emerged and are beginning to reshape the global value chains with disruption innovations, often undermining established labor rights. This is evident in developing countries where minimal labor market regulation, poor education quality, and low social mobility opportunities lead to large informal economies. The informal economy—a diversified set of economic activities, enterprises, jobs, and workers that are not regulated or protected—is a fast-growing sector that includes the world's poorest and most vulnerable individuals.

With global unemployment projected at 207 million in 2022, surpassing its 2019 level by some 21 million, the United Nations International Labor Organization (ILO) estimates that 61% of all workers (an estimated two billion people) are employed in the informal economy[4] (Office, 2022), and 94% of the world's informal employment is in emerging and developing countries. In this sense, the informal economy represents the foundation of the world's economic development, with informal economy SMEs contributing to more than 50% of most countries' gross domestic product.

In the initial stages of the Covid pandemic, informal employees were three times more likely than formal employees to lose their jobs. Millions of home-based workers in the garment industry saw their orders grind to a halt, with brands refusing to pay for work already completed. At the same time, they had no safety net of health care, income, or social protections, forcing many to resort to asset-depleting coping mechanisms (Ford Foundation, 2021). As the pandemic evolved, formal wage workers returned to employment. In contrast, informal waged work has remained below its pre-crisis level in a sample of ten middle-income countries. The UN ILO suggests that formal enterprises have managed to weather the crisis better than informal ones.

Labor unions face several challenges caused by the increasing digitalization of work and workers. Some of these challenges are political and strategic; for example, how to address the lack of rapidly changing employment relationship dynamic, identifying and recommending solutions to gaps within current social legislation, and collective bargaining in a digital age free of geographical constraints. Other challenges relate to unions' internal communications, transparency and trust, and better union decision-making and operations efficacy. In either case, unions are beginning to negotiate for much more vital workers' rights regarding the increasing adoption of IR4.0 technologies in workplaces. This has a clear impact on the power of workers to negotiate decent working conditions. Organized labor is finding it challenging to keep up with the current risks as the pace of informal, precarious, and outsourced work has been intensifying. This would appear to suggest that workers' power is in decline (Dirksen & Herberg, 2021).

In many world regions, digital technologies are being contested by workers and organized labor, giving rise to offensive and defensive labor struggles. At the same time, more and more digital innovations leveraging IR4.0 technologies are being explored by cooperative associations and the representation of workers, especially in the platform economy.

The conditions for workers may already be negatively affected, and labor unions have the arguments to demand better protection for workers:

- Collective bargaining is for all workers, regardless of employment status.
- Mental and physical health issues associated with automation technologies decrease disaggregated completion times while increasing process complexity.
- Discrimination or bias applied to upskilling/training opportunities.
- Deskilling, economic insecurity, less mobility, and ultimately job loss.
- Privacy issues related to each worker's rights to agree to, block, or amend individual data collection and define the purposes/limitations of using advanced analytic tools and their extracted insights. Information on algorithmic systems should be a part of an onboarding process for new workers. Unions must explore the potential value of leveraging digital technologies for their purpose; for example, by allowing digital labor unions to have open access to employer-worker data or to collect their private worker data. While responsibly collecting worker data, unions can better identify issues and target work-change arguments with evidential case studies of the real impact of IR4.0 technologies on workers. All unions can also better identify the benefits of using IR4.0 technologies as a tool to understand, communicate with, and organize workers. The Century Foundation, an independent think tank that conducts research that drives policy innovations that make people's lives better, has suggested the following:
  1. Globally, the older generation of workers are leaving employment, and the younger generation, who depend on digital platforms for communication and information, is replacing them. Union leaders must reach this new technology-savvy generation when, where and how they communicate, or risk not reaching them at all.
  2. Digitalization has disrupted the paradigm of worker organization, creating new opportunities by connecting traditional tactics with new technologies.

3. Digital tools make it easier and smoother for workplaces to organize. Digital unionism unions sidestep the difficulties of accessing employees at anti-union workplaces (Walker, 2017).

Digital tools are essential in successful union organization in workplace drives and strikes. When it comes to organizing a union in a workplace, these tools help associations map the workplace, communicate anonymously with workers, and build unity among employees (Zuckerman et al., 2015).

By leveraging social media platforms, discussion forums, real-time reporting, digital/virtual union representatives, and other networking platforms, workers' conditions and rights can be broadly disseminated. This can also mean that workers receive faster communication, more transparency in the decision-making process, and more involvement in designing new digital union services. Yet the most significant opportunity may be in the informal sector. The low-skilled workers within the informal economy will be the hardest hit by automation and other IR4.0 technology adoption. It is challenging for labor unions to organize the informal economy because of the fundamental differences and constraints between informal and formal economy workers. Moreover, the organization and objectives of labor unions do not permit a simple extension of their traditional activities to cover informal economy issues.

On the other hand, it is an opportunity because the informal economy may motivate renewing the labor union movement by expanding membership and collective bargaining coverage to protect labor rights, strengthen workers' collective voice, and influence social and economic policies.

Digital platforms to better organize the informal sector have introduced innovations. For example, the Lynk platform, which was acquired by Eden Life, Africa's first home concierge services scheduling platform in April 2022, connects businesses and households with verified domestic workers, artisans, and blue-collar workers in Nairobi. By September 2018 the platform had successfully "lynked" 20,000 jobs. There are currently over 283 platforms in Africa that provide various services (Hunter et al., 2018).

In a November 2021 press release, and in partnership with Women in Informal Employment: Globalizing and Organizing, the Ford Foundation announced a five-year, $25 million grant to women-led informal worker networks to support a global movement calling on governments to invest in protection for informal workers (Ford Foundation, 2021).

As more policymakers, governments, and organizations realize the importance of the gig economy, supporting and introducing digital labor platforms becomes an economic development strategy that can locate jobs in places that need them. Still, it is essential to understand how this might influence workers' livelihoods and within the communities where these new digital workers may live.

Union leaders have the opportunity, and the responsibility, to take digitization step by step to build a solid digital presence and allow their unions to flourish into the future.

THERE IS A FUTURE FOR DIGITAL LABOR UNIONS

The labor union movement certainly has a future, but it will involve changes. Lots of changes. Workers and managers alike are breaking ranks with those who push the notion that IR4.0 technologies are equitable and inclusive. Informal workers are becoming more technologically savvy and are discovering new networking models that coordinate collective action on a global scale. As the negative impacts of IR4.0 technologies outweigh the positive effects even slightly, workers' resistance will grow as they try to rebalance the power dynamic. To survive and succeed, digital unions must be strategically aligned to support workers who want to be included in the digital transformation and are willing to change responsibly.

Most, if not all, labor markets will adapt and adopt digital technologies in the future. Still, they would want a voice in the regulation, governance, and implementation of IR4.0 technologies to ensure a fair balance between social and economic benefits. However, getting from here to there will require fully understanding the complexities of IR4.0 technologies, how they will converge to create an even more advanced breed of technologies (IR5.0?), and, finally, how to leverage their immense opportunities for work, workers, and the workplace.

NOTES

1 Investors increasingly include these non-financial factors in their analysis process to identify material risks and growth opportunities.
2 UN SDGs, also known as the Global Goals, are 17 goals with 169 targets that all UN member states have agreed to achieve by 2030. It is a vision for an inclusive and more

equitable policy environment and coordinated investment initiatives to promote sustainable development in all member countries.

3 The Amazon Labor Union is an independent, worker-led, democratic labor union founded by Amazon workers in Staten Island, New York. The union was formed in April 2021 by a group of concerned workers led by ALU President Chris Smalls, a Process Assistant at JFK8 in Staten Island who was fired by Amazon management for organizing protests over Amazon's unsafe COVID-19 protocols. www.amazonlaborunion.org/

4 Here, the "informal sector" refers to self-financed, under-capitalized, small-scale, unskilled labor-intensive production. An alternative definition is a "process of income generation" that is "unregulated by the institutions of society, in a legal and social environment in which similar activities are regulated" (Pratap & Quintin, 2006). "The Informal Sector in Developing Countries: Output, Assets and Employment," WIDER Working Paper Series RP2006-130, World Institute for Development Economics Research (UNU-WIDER).

REFERENCES

Bai, C., Dallasega, P., Orzes, G., & Sarkis, J. (2020). Industry 4.0 technologies assessment: A sustainability perspective. *International Journal of Production Economics*, *229*, 107776. https://doi.org/10.1016/j.ijpe.2020.107776

Blayone, T. J. B., & VanOostveen, R. (2021). Prepared for work in Industry 4.0? Modelling the target activity system and five dimensions of worker readiness. *International Journal of Computer Integrated Manufacturing*, *34*(1), 1–19. https://doi.org/10.1080/09511 92X.2020.1836677

Cabinet Office, C., Government of Japan. (n.d.). *Society 5.0*. Retrieved April 25, 2022, from www8.cao.go.jp/cstp/english/society5_0/index.html

Carpi, R., Littmann, A., & Schmitz, C. (2019, June 2). *The future of manufacturing: Your people.* www.mckinsey.com/business-functions/operations/our-insights/the-future-of-manufacturing-your-people

Clark, D. (2021, September 30). *Global SMEs 2020*. Statista. www.statista.com/statistics/1261 592/global-smes/

Coherent Market Insights. (2021). *IT Robotic Automation Market Analysis* (No. CMI845; p. 130). www.coherentmarketinsights.com/market-insight/it-robotic-automation-market-845

Dirksen, U., & Herberg, M. (Eds.). (2021). *Trade Unions in Transformation 4.0: Stories of Unions Confronting the New World of Work.* Friedrich-Ebert-Stiftung, Global Policy and Development. http://library.fes.de/pdf-files/iez/17798-20210602.pdf

Ford Foundation. (2021, November 16). *Ford Foundation announces five-year, $25 million grant to women-led global worker networks calling for a just economic recovery for the world's 2.1 billion informal workers.* Ford Foundation. www.fordfoundation.org/news-and-stories/news-and-press/news/ford-foundation-announces-five-year-25-mill ion-grant-to-women-led-global-worker-networks-calling-for-a-just-economic-recov ery-for-the-world-s-21-billion-informal-workers/

Gardoni, P. (2020). *From Internet of Things to System of Systems: Market analysis, achievements, positioning and future vision of the ECS community on IoT and SoS.* www.eurotech.com/ attachment/download?id=2094

Gartner. (2022, March 1). *Hybrid workforce: Rethink work to better support on-site, contingent and remote workers.* Gartner. www.gartner.com/en/human-resources/insights/manag ing-hybrid-workforce

Gillis, A. (2022, March). *What is IoT (Internet of Things) and How Does it Work?* IoT Agenda. www.techtarget.com/iotagenda/definition/Internet-of-Things-IoT

Hunter, R., Chernay, J., & Mothobi, O. (2018, December). *African digital platforms and the future of financial services.* http://researchictafrica.net/wp/wp-content/uploads/2018/12/DInfo_V11.pdf

Laplante, P. A., Kassab, M., Laplante, N. L., & Voas, J. M. (2018). Building caring healthcare systems in the Internet of Things. *IEEE Systems Journal, 12*(3), 3030–3037. https://doi.org/10.1109/JSYST.2017.2662602

Lund, S., Madgavkar, A., Manyika, J., Smit, S., Ellingrud, K., Meaney, M., & Robinson, O. (2021). The postpandemic economy: The future of work after COVID-19. *McKinsey Global Institute,* 152.

Manyika, J., Lund, S., Chui, M., Bughin, J., Woetzel, J., Barta, P., Ko, R., & Sanghvi, S. (2017). *Jobs Lost, Jobs Gained: Workforce Transitions in A Time of Automation.* McKinsey Global Institute.

MEXT. (2018). *Human resource development for Society 5.0.* www.mext.go.jp/b_menu/activity/detail/pdf2018/20180605_001.pdf

Middleton, P., Tully, J., & Kjeldsen, P. (2013, November 18). *Forecast: The Internet of Things, Worldwide, 2013.* Gartner. www.gartner.com/en/documents/2625419

Neal, B., & Caplan, S. (2018, May 15). *Leading the way to the future of work.* PwC. www.pwc.com.au/digitalpulse/intelligent-digital-leading-future-work.html

Office, I. L. (2022). *World employment and social outlook trends 2022.* International Labour Organization (ILO). https://public.ebookcentral.proquest.com/choice/PublicFullRecord.aspx?p=6941163

Osservatorio Transizione Industria 4.0: La Ricerca 2020–2021. (2020). www.osservatori.net/it/ricerche/osservatori-attivi/transizione-industria-40

Pratap, S., & Quintin, E. (2006). Are labor markets segmented in developing countries? A semiparametric approach. *European Economic Review, 50*(7), 1817–1841. https://doi.org/10.1016/j.euroecorev.2005.06.004

Singapore Economic Development Board (EDB). (2019a). *Smart Industry Readiness Index | The prioritisation matrix catalysing the transformation of manufacturing.* www.edb.gov.sg/content/dam/edb-en/about-edb/media-releases/news/the-smart-industry-readiness-index/Smart-Industry-Readiness-Index-Prioritisation-Matrix-Whitepaper.pdf

Singapore Economic Development Board (EDB). (2019b, October 22). *The Smart Industry Readiness Index.* www.edb.gov.sg/en/about-edb/media-releases-publications/advanced-manufacturing-release.html

Walker, M. (2017, June 13). *Three strategies unions are considering for their survival.* The Conversation. http://theconversation.com/three-strategies-unions-are-considering-for-their-survival-78992

Watson, M. D., Mesmer, B. L., & Farrington, P. A. (2020). *Engineering Elegant Systems: Theory of Systems Engineering.* NASA. www.nasa.gov/sites/default/files/atoms/files/nasa_tp_20205003644_interactive2.pdf

Zinser, M., Rose, J., & Sirkin, H. (2015, August 23). *The robotics revolution: The next great leap in manufacturing.* BCG Global. www.bcg.com/publications/2015/lean-manufacturing-innovation-robotics-revolution-next-great-leap-manufacturing

Zuckerman, M., Kahlenberg, R. D., & Marvit, M. Z. (2015). Virtual labor organizing: Could technology help reduce income inequality? *The Century Foundation | Tcf.Org,* 10.

7

Digital Stress and Coping

Christiane Hagmann-Steinbach
Hagmann-Steinbach Consulting, Stuttgart, Germany

CONTENTS

STRESS – WHAT IS IT ACTUALLY?

Stress is not a new phenomenon. Hans Selye, a Hungarian-Canadian doctor, described stress as far back as "in the 1930s and 1940s" (O'Donnell, 1976, p. 600). He is considered the "father of stress," although not entirely uncontroversial today because of Big Tobacco's sponsorship of research projects (Petticrew & Kelley, 2011, p. 411). The American Psychological Association defines stress as "… physiological or psychological response to internal or external stressors. Stress involves changes affecting nearly every system of the body, influencing how people feel and behave" (APA, n.d.). Typical stress conditions are by no means limited to big, rare life events like accidents, illness, the loss of a family member, or war (Ernst et al., 2022). At work there are often less severe stimuli that individuals find stressing and that cause stress; for example, time pressure, a critical presentation, a brand-new job, a negative working atmosphere, as well the daily hassle like traffic jams on the way to the office and many more (Ernst et al., 2022). Experience in companies shows that those conditions do not trigger stress in the same way in every

DOI: 10.1201/9781003371397-7

FIGURE 7.1
Stressed person in the office.

person. But in the working life every employee must deal with demands that either exceed their capabilities and coping options or do not, as the following figure shows (Strobel & v. Krause, 1997 *in* Stadler, 2006*)*.

When a person perceives a difficult and challenging situation as manageable, even exhilarating, many experts speak of eustress (Goal, 2021; Medical News Today, 2022). This is the positive side of stress, which is not the subject here. This article focuses on the other side of the coin, the distress that is perceived as burdensome (Goal, 2021). In the following, stress is defined as a negative consequence of a discrepancy between the stresses acting on a person and their individually perceived coping possibilities (Stadler, 2006). That means stress at work is a sense of not being capable of addressing the specific demands, needs, and events.

The physical reaction is always the same: When individuals are in danger, the body responds by releasing hormones to muster all its strength and to supply energy (Techniker Krankenkasse, 2017); there is a higher pulse, faster breathing, slower stomach and intestine activity, more blood glucose, dampened immune system, so that the person can fight or run away (Medical

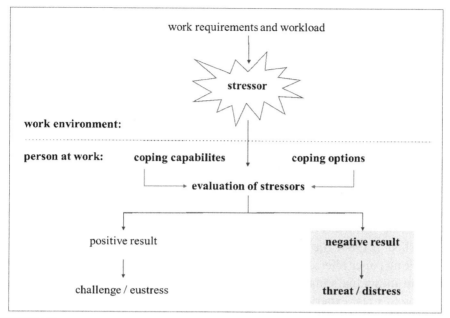

FIGURE 7.2
Stress as negative strain or distress (own illustration, following Stadler, 2006).

News Today, 2022). These responses are all designed to keep the person alive in a dangerous situation (Techniker Krankenkasse, 2017). However, if stress is persistent, the consequences can be harmful (Medical News Today, 2022). Severe stress causes mind-body changes that may severely affect mental and physical health. Among the effects and diseases that can arise from this are loss of concentration, fatigue, cardiovascular disturbances, muscle cramps, headaches, inflammation of the gastrointestinal tract, skin eczemas, anxiety, burnout, insomnia, and depression (AIS, n.d.).

DIGITAL STRESS CAN MAKE YOU ILL AND IS POTENTIALLY VERY EXPENSIVE

In 2016 a German study on the impact of digitization at work highlighted several benefits for the employees (Arnold et al., 2016): 29% of the respondents experienced physical relief, 56% a better work performance and 32% appreciated a higher decision-making scope (Arnold et al., 2016). The results of a practitioner survey at a consulting customer in the service industry confirm the positive picture: The individuals very much appreciate that communication by email is fast and uncomplicated. They are pleased that they can meet customers without having to travel, thanks to video conferencing, and they benefit from document sharing.

However, the research report from Arnold and his colleagues refers to specific downsides (Arnold et al., 2016): 78% of the respondents say that technological change requires the steady development of skills and competencies, while 65% feel that there is work densification with a flood of information that is difficult to cope with. If in these cases individuals are unable to fulfill their responsibilities, digital stress can arise.

The fact that dealing with digital technologies can trigger stress is nothing new. Craig Brod was one of the first to write about such effects in his book published in 1984: *The Human Cost of the Computer Revolution*. This is what a working paper refers to in a definition of digital stress (Gimpel et al., 2018): It is an individual's inability to deal with new technology in a healthy way, leading to stress experiences.

Digital stress can affect the well-being and the quality of life (Gimpel et al., 2019). In the short term it leads to greater emotional and physical exhaustion (Gimpel et al., 2019; Riedl et al., 2020). Digital stress also triggers cognitive and emotional irritation, a "subjectively perceived… strain in context of the working environment" (Mohr et al., 2005, p. 44). Irritation is a state between "mental fatigue and mental illness" (Mohr & Rigotti, 2014, p. 1). Cognitive irritation manifests itself, for example, in an individual's inability to switch off from work.

Other health problems that arise from digital stress include headaches, nighttime sleep disorder, and general fatigue (Gimpel et al., 2018). Gimpel's 2019 study with 5,000 employees in Germany cites musculoskeletal disorders most frequently, followed by cardiovascular disease, mental impairment, and neurological-sensory disorders. Digestive complaints occur the least frequently (Gimpel et al., 2019). An interesting aspect of this study is the following: A differentiation between subjects who experienced strong digital stress and those who had only a slight sense of stress. It turned out that 39% of those who experienced strong levels of digital stress suffered from mental health problems. Thus, psychological impairments take second place behind musculoskeletal disorders. For the other illnesses, the difference between the two groups was only between just under 13% and 15%.

Being under digital stress is not only an issue for the individual but also for organizations. Digital stress translates into absenteeism (Tarafdar et al.,

2007). It also negatively impacts job performance and productivity (Tarafdar et al., 2007). Furthermore, individuals who experience digital stress are more likely to be dissatisfied with their work and express lower user satisfaction with systems (Riedl et al., 2020). Attachment to employers decreases, and employees are more likely to think about changing jobs or careers (Gimpel et al., 2019; Ragu-Nathan et al., 2008).

When interpreting the results, it should be noted that people often do not consciously perceive stress even though stress hormones have already risen and parameters of the cardiovascular system such as blood pressure are deteriorating. Thus, stress often exists before people realize they are stressed (Riedl et al., 2020). It should also be noted that digital stress increases general work stress (Riedl et al., 2020).

It may be valuable to investigate the economic impact of digital stress. Unfortunately, there is currently a lack of published data available. However, there are numbers on workplace stress that may give an indication. It is widespread and takes an enormous toll on the economy. "Around half of European workers consider stress to be common in their workplace, and it contributes to around half of all lost working days" (EU-OSHA, n.d.). "… the cost to Europe of work-related depression was estimated to be €617 billion annually" (Hassard et al., 2014, p. 7). In the U.S. the situation is similar. The American Institute of Stress refers to numerous studies and points out "… that job stress is far and away from the major source of stress for American adults and that it has escalated progressively over the past few decades" (Boyd, 2022). An estimated one million workers are absent every day because of stress, which costs the U.S. industry more than $300 billion annually (Heckman, 2021).

In short, digital stress has a negative impact on individuals and companies. To counteract this, measures are necessary. However, this requires a good understanding of the screws to turn.

DIGITAL STRESS – MANY ROOT CAUSES

The reasons for stress are multidimensional. Several researchers and practitioners have published on this topic (Gimpel et al., 2018; Nisafani et al., 2020; Ragu-Nathan et al., 2008; Riedel, 2020; Tarafdar et al., 2007). Literature shows that many factors can create digital stress. In their research summarizing different studies, Nisafani and colleagues classified the factors that cause digital stress into two categories (Nisafani et al., 2020): Not surprisingly, the performance of the ICT system is one of them. The other category includes performance-related factors.

"System breakdown…" is a typical system performance factor (Nisafani et al., 2020, p. 247). This comprises malfunctions of the systems (Nisafani et al., 2020). In a study on digital stress in Germany, the unreliability of digital technologies and media is the third-most stressor (Gimpel, et al., 2019). To give an example, the last thing users want when they are in the middle of a video call with an important customer is a system crash that costs a lot of time to fix.

Long response time and lack of technical support are other factors that fall into this group (Riedl et al., 2020). If users are unable to use the functionalities offered by the system or if the system is not easy to use, "usability issues" may occur and cause digital stress (Nisafani et al., 2020, p. 247). Finally, "security issues" fall into this category (Nisafani et al., 2020). This comprises any malicious cyber threats and attacks that seek to disrupt digital operations, steal, or damage data. Riedel points to another cause of stress: Missing functions that are needed for task performance and superfluous functions that the systems have but that no one needs (Riedl et al., 2020).

Many of the non-system performance-related factors are closely related to the individual and the user behavior (Nisafani et al., 2020).

The high complexity of digital technologies is a stressor that researchers have identified (Gimpel et al., 2019; Ragu-Nathan et al., 2008; Riedl et al., 2020). Users feel overwhelmed if their abilities are not sufficient to use the ICT or if the functionality is difficult to understand. They must learn how to use and control the new systems. This may require a high learning effort. As technologies "change rapidly" in the ICT industry, users have to constantly adapt to new requirements, which can "take months to learn" (Ragu-Nathan et al., 2008, p. 422).

An important stress-producing condition is the lack of work-life balance. Employees can be reached "anywhere and anytime" (Ragu-Nathan et al., 2008, p. 421) . The "invasion" of digital technologies is making the distinction between personal life and work more difficult (Ragu-Nathan et al., 2008, p. 426). It is a kind of communication pressure if individuals "… feel forced to…" be always available online, be it at home or in the office, or if they have to answer immediately (Ragu-Nathan et al., 2008, p. 421). As a matter of fact, the constant connectivity leads to a feeling of longer working hours. This can have a negative effect on job satisfaction (Ragu-Nathan et al., 2008). As experience shows, a reason for this behavior is still widespread in companies;

the fear of not being able to lose connection with the online business world, in other words FOMO, fear of missing out.

Another stressor is "information and communication overload" (Ragu-Nathan et al., 2008, p. 421). It seems plausible that individuals at work often must handle more information than they can cope with (Ragu-Nathan et al., 2008). In addition to that, the increasing number of ICT tools can lead to "excessive multitasking," which is, as research shows, a stressor as well (Ragu-Nathan et al., 2008, p. 422).

At a customer's site, the employees may work with laptops, phones and smart phones, a management information and enterprise planning resource system, an office suite with word processor, spreadsheet editor, presentation program, personal information manager, an image processing app, a ticketing system for assistant services, the frontend of a content management system for website editing and newsletter content creation, a virtual collaboration app for projects, different whiteboard solutions, a video communication app for online meetings with external stakeholders, and there are other apps in the pipeline. A study from 2019 shows that, on average, German employees used at least 11 different technologies and media a week (Gimpel et al., 2019). Those who have many digital tools but seldom use them feel more stress than those who have few digital tools and use them often (Gimpel et al., 2019).

Many employees have the feeling that ICT forces them to "work faster" (Ragu-Nathan et al., 2008, p. 421). The expected constant changes in digital technology makes knowledge obsolete and leads to so-called techno-uncertainty (Tarafdar et al., 2007). Fast changes through technological progress may lead to insecurity and fear of losing the job (Gimpel et al., 2019; Riedl et al., 2020).

Performance monitoring via digital technologies is another potential stressor. The respondents in the study by Gimpel and colleagues rate it as the strongest stress factor. The electronic performance monitoring made possible by digital technologies ranks second in the investigation. (Gimpel et al., 2019). Riedl confirms in his research that fear of personal data theft is among the top stressors (Riedl et al., 2020). This applies to the case when information is easily accessible to unauthorized persons or when personal data can be stolen (Gimpel et al., 2019).

Furthermore, digital stress increases general work stress. Reducing digital stress, therefore, leads to less work stress and protects health (Riedl et al., 2020).

WAYS TO COPE WITH DIGITAL STRESS

Undoubtedly it is a challenge to keep away from or fight digital stress. But there are several things that can be done. According to the research of Gimpel and colleagues, fortunately many users become active themselves. The most common coping strategies applied by employees are the following: Staying positive, taking steps to improve the situation, using humor, creating a plan, and living with it (Gimpel et al., 2019). People with "technology self-efficacy" (Nisafani et al., 2020, p. 251) who are active in managing digital stress rate their health better than those who do little about it (Gimpel et al., 2019). But it is not enough to leave the initiative to the individual.

On the organizational level, measures include "technology-related inhibitors," as described in a study that summaries numerous research findings (Nisafani et al., 2020, p. 251):

- Reliability, meaning an ICT system operating "in a defined environment without failure" (ASQ, n.d.) is taken into consideration as helpful. It diminishes overload.
- User experience with other systems reduces complexity as own experience shows.
- Usefulness contributes to the avoidance of trouble.
- Usability of the digital solution. There are plenty of strategies to determine it. "A quick and dirty usability scale" is the System Usability Scale, a ten-question assessment tool, that people fill out after the usage of an ICT system (Brooke, 1995; General Services Administration, n.d.). A high score comes with low-stress potential.

What definitely should be considered, is the workplace equipment (Gimpel et al., 2019). Consulting experience shows that it is not enough to provide employees with minimal equipment such as a standard desktop pc. To be able to do their job professionally and without digital stress, many individuals also need more; for example, a large screen, a separate camera, and a high-quality headset. In connection with this, it is important to consider that technical help provision through a well-organized helpdesk is vital for stress reduction. This has a positive impact on user satisfaction and productivity (Riedel, 2020).

Also to be taken into account is the design of the work. This includes the activity itself, as well as "the work environment, the work organization, the workplace", and the "work equipment" (Gimpel, et al., 2019, p. 38). This also means, for example, that it is often insufficient to provide employees who work remotely with a company laptop.

A "non-technology-related inhibitor" of digital stress is worker involvement in digitization (Nisafani et al., 2020). As practice shows, participation in problem identification and solution leads not only to sound solutions but also to higher acceptance rates. A good example comes from a client who planned to install virtual call centers in pre-pandemic times. In this business the agents handle queries and make calls on behalf of their current or potential clients. Hence, one of the challenges was to familiarize newly employed agents with the use of various customer management systems they were working with. In a participatory process based on an agile scrum development methodology, an employee team developed numerous measures to make the onboarding as easy as possible. One successful approach was a prototype (the so-called "home-office simulator") with the help of which the newcomers could practice different customer management systems. A "remote work bible" contained advice on the technical equipment requirements and remote help desk contacts but also break regulations and rules for communication with digital media such as emails and chats. In this case, the deep involvement of the users in the digitization process from the beginning to the end led to high acceptance of the measures.

As described in a study, the client confirmed that training on new systems enabled them to avoid digital stress (Ragu-Nathan et al., 2008). Furthermore, the agent team asked their management to offer IT training at a low level, early on, and repeatedly. They additionally recommended supplying volunteer multipliers who are acquainted with the new digital technologies as points of contact for employees.

One component that should not be underestimated is the support provided by the respective manager. A good relationship with one's supervisor can reduce digital stress (Gimpel et al., 2019). Likewise, it can prevent employees' tendency to go to work sick (Böhm et al., 2016). The negative effects of digital stress can also be reduced if employees can flexibly arrange their working hours and locations (Böhm et al., 2016).

And if digital stress does make itself felt, that is no reason to despair. Many actors such as the World Health Organization, health insurance companies, or self-help organizations offer free material to address general and work stress (AOK, n.d; Kleinschmidt, 2017; WHO, 2020): It includes advice, contact persons, as well as questionnaires to analyze the personal stress situation and diaries to identify precisely when the individual feels stressed, what causes the stress, and how it affects work. Stress management techniques, e-coaching, and training to relax and maintain a more balanced life are part of such programs. They help employees become more resilient and better cope with digital stress. As one company has just experienced, and as a study confirms, it can also be helpful to assess the digitization situation in the organization; for example, with the help of an employee survey (Böhm et al., 2016). This can be a good starting point for identifying stress factors and eliminating actions.

SUMMARY

Research and practitioner experience show that digitization can trigger digital stress, which negatively affects the well-being and, in the longer term, the health of employees. Work stress is also associated with considerable damage to businesses and the economy. Companies should therefore become aware of this issue and actively address it within their sphere of influence. At the same time, science must continue to explore how the various opportunities of digitization can be used in the interest of safe, healthy and good work (Rothe et al., 2019).

REFERENCES

American Institute of Stress. (n.d.). *How stress affects your body*. Retrieved June 3, 2022, www.stress.org/how-stress-affects-your-body

American Psychological Association. (n.d.). *APA Dictionary of Psychology. Washington*. Retrieved May 9, 2022, https://dictionary.apa.org/stress

American Society for Quality. (n.d.). *What is reliability? Milwaukee*. Retrieved June 3, 2022, https://asq.org/quality-resources/reliability

AOK. (n.d.). Stress im Griff [Stress under control]. Retrieved May 20, 2022, from www.stress-im-griff.de/

Arnold, D., Butschek, S., Steffes, S., & Müller, D. (2016). *Monitor – Digitalisierung am Arbeitsplatz: Aktuelle Ergebnisse einer Betriebs- und Beschäftigtenbefragung*. Berlin: Bundesministerium für Arbeit und Soziales.

Böhm, S., Bourovoi, K., Brzykcy, A., Kreissner, L., & Breier, C. (2016). *Auswirkungen der Digitalisierung auf die Gesundheit von Berufstätigen: Eine bevölkerungsrepräsentative Studie in der Bundesrepublik Deutschland*. U. S. Gallen. St. Gallen. Retrieved May 5, 2022, from www.alexandria.unisg.ch/252056/13/20191220_Studie_Digitalisierung_Gesundheit_Final.pdf

Boyd, D. (2022, September 22). Workplace stress. *The American Institute of Stress*. www.stress. org/workplace-stress

Brooke, J. (1995, 11). *SUS: A quick and dirty usability scale*. Reading. Retrieved May 3, 2022, www.researchgate.net/profile/John-Brooke-6/publication/228593520_SUS_A_quick_ and_dirty_usability_scale/links/5f24381392851cd302cbaf25/SUS-A-quick-and-dirty-usability-scale.pdf?origin=publication_detail

Ernst, G., Franke, A., & Frankowiak, P. (2022). *Stress und Stressbewältigung*. Bundeszentrale für gesundheitliche Aufklärung. Köln. doi:10.17623/BZGA:Q4-i118-2.0

European Agency for Safety and Health at Work. (n.d.). *European Agency for Safety and Health at Work*. Retrieved April 19, 2022, https://osha.europa.eu/en/themes/psychosocial-risks-and-stress

General Services Administration. (n.d.). *System Usability Scale /Post Test Questions*. D. A. Administration. Retrieved June 3, 2022, https://digital.gov/resources/digitalgov-user-experience-resources/digitalgov-user-experience-program-usability-starter-kit

Gimpel, H., Lanzl, J., Manner-Romberg, T., & Nüske, N. (2018). *Digitaler Stress in Deutschland: Eine Befragung von Erwerbstätigen zu Befragung von Erwerbstätigen zu Belastung und Banspruchung durch Arbeit mit digitalen Technologien*. Düsseldorf: Hans-Böckler-Stiftung. Retrieved May 5, 2022, www.boeckler.de/pdf/p_fofoe_WP_101_2018.pdf

Gimpel, H., Lanzl, J., Regal, C., Urbach, N., Wischniewski, S., Tegtmeier, P., ...Derra, N. (2019). *Gesund digital arbeiten?! Eine Studie zu digitalem Stress in Deutschland*. P. W. FIT. Augsburg. https://doi.org/10.24406/fit-n-562039

Goal, S. (2021). *Eustress und Distress: Psychische Belastung hat verschiedene Gesichter*. AOK Baden-Württemberg. Stuttgart. Retrieved June 4, 2022, www.aok.de/bw-gesundnah/psyche-und-seele/eustress-und-distress-teufelchen-und-engelchen

Hassard, J., Teoh, K., Cox, T., Dewe, P., Cosmar, M., Van den Broek, K., Gründler, R., & Flemming, D. (2014). Calculating the costs of work-related stress and psychosocial risks: Literature review. *Publications Office*. https://op.europa.eu/s/w8Xl

Heckman, W. (2021, April 9). Worrying workplace stress statistics. *The American Institute of Stress*. www.stress.org/42-worrying-workplace-stress-statistics-2

Johnson, J. (2021). *Number of internet users worldwide from 2005 to 2021*. Statista. Hamburg. Retrieved June 3, 2022, www.statista.com/statistics/273018/number-of-internet-users-worldwide/

Kleinschmidt, C. (2017). Kein Stress mit dem Stress [No stress with stress]. *Initiative Neue Qualität der Arbeit*. www.psyga.info

Landesmedienzentrum Baden-Württemberg. (n.d.). *Geschichte des Computers, Die Geschichtes des Computers von den Anfängen bis heute*. Karlsruhe. Retrieved May 11, 2022, www.lmz-bw.de/medien-und-bildung/medienwissen/informatik-robotik/historisches/geschichte-des-computers/

Medical News Today. (2022). Eustress vs. distress: What is the difference? *Medical News Today*. Brighton. Retrieved June 3, 2022, www.medicalnewstoday.com/articles/eustress-vs-distress#definitions

Mohr, G., & Rigotti, T. (2014). Irritation (Gereiztheit). GESIS – Leibniz-Institut für Sozialwissenschaften e.V. GESIS – Leibniz-Institut für Sozialwissenschaften, Zusammenstellung sozialwissenschaftlicher Items und Skalen (ZIS), Mannheim. Retrieved June 3, 2022, https://zis.gesis.org/kategorie/showZISWithGesisSearch?source={%22query%22:{%22bool%22:{%22must%22:[{%22query_string%22:{%22query%22:%22IrrITATION%22,%22default_operator%22:%22AND%22}}],%22filter%22:[{%22term%22:{%22type%22:%22zis_scales%22}}]}}}

Mohr, G., Rigotti, T., & Müller, A. (2005). Irritation - ein Instrument zur Erfassung psychischer Befindensbeeinträchtigungen im Arbeitskontext. Skalen- und Itemparameter aus 15 Studien. *Zeitschrift für Arbeits – und Organisationspsychologie*, 49, 44–48. Göttingen: Hogrefe Verlag. Retrieved May 28, 2022, www.researchgate.net/publ ication/260551925_Irritation_-_ein_Instrument_zur_Erfassung_psychischer_ Befindensbeeintrachtigungen_im_Arbeitskontext_Skalen-_und_Itemparameter_aus_ 15_Studien

Nisafani, A. S., Gaye, K., & Mahony, C. (2020). *Workers' technostress: a review of its causes, strains, inhibitors, and impacts.* https://doi.org/10.1080/12460125.2020.1796286

O'Donnell, F. J. (1976). Stress without distress (book review). *67(4). British Journal of Psychology.*

Petticrew, M., & Kelley, L. (2011). *The "Father of Stress" Meets "Big Tobacco": Hans Selye and the Tobacco Industry. American Public Health Association.* 101(3), 411–418. Retrieved June 4, 2022, https://ajph.aphapublications.org/doi/epub/10.2105/AJPH.2009.177634

Ragu-Nathan, T., Tarafadar, M., Ragu-Nathan, B., & Tu, Q. (2008). The consequences of technostress for end users in organizations: Conceptual development and empirical validation. *Information System Research*, 19(4). Retrieved May 5, 2022, www.researchg ate.net/publication/220079808_The_Consequences_of_Technostress_for_End_Users_ in_Organizations_Conceptual_Development_and_Empirical_Validation

Riedel, R. (2020). *Digitaler Stress Wie er uns kaputt macht und was wir dagegen tun können.* Wien: Linde Verla GmbH.

Riedl, R., Fischer, T., Kalischko, T., & Reuter, M. (2020). *Digitaler Stress Eine Befragungsstudie im deutschsprachigen Raum.* Fachhochschule Oberösterreich Campus Steyr, Johannes Kepler Univertität Linz, & Friedrich Wilhelms-Universität Bonn. Steyr. Retrieved May 4, 2022, https://forschung.fh-ooe.at/digitaler-stress-studie/

Rothe, I., Wischniewski, D. S., Tegtmeier, D. P., & Tisch, D. A. (2019). Arbeiten in der digitalen Transformation – Chancen und Risiken für die menschengerechte Arbeitsgestaltung. *Zeitschrift für Arbeitswissenschaft, 3.* https://creativecommons.org/licenses/by/4.0/ deed.de

Stadler, P. (2006). Psychische Belastungen am Arbeitsplatz - Ursachen, Folgen und Handlungsfelder der Prävention. B. L. Lebensmittelsicherheit. Retrieved May 5, 2021.

Tarafdar, M., Tu, Q., Ragu-Nathan, B., & Ragu-Nathan, T. (2007). The impact of technostress on role stress and productivity. *Journal of Management Information Systems, 24*(1), 303–334. L. Taylor & Francis. DOI 10.2753/MIS0742-1222240109

Techniker Krankenkasse. (2017). *Stress Belastungen besser bewältigen.* Techniker Krankenkasse. Hamburg. Retrieved June, 2022, www.tk.de/resource/blob/2023234/5535b9478a9be 8fcabb0a1c6ea7f677e/tk-broschuere-stress-data.pdf

WHO. (2020). Doing what matters in times of stress: An illustrated guide. World Health Organization. https://apps.who.int/iris/handle/10665/331901

8

Economics of Mind, Body, Spirit – Experimental Games Reveal New Human Choices

Syed Shurid Khan

Asian Institute of Technology (AIT) | School of Management, Bangkok, Thailand

CONTENTS

HUMAN RATIONALITY AND MOTIVATION

What motivates humans and influences their choices and decision-making are still a mystery and a focus of curiosity in various fields of science, including psychology, neuroscience, and economics, to name a few. Scientists and related think tanks are looking for answers to some centuries-old questions, such as how the human mind works, which is a particularly important question in today's world that mainly relies upon intellectual capability rather than muscle power. In this age of data, do individuals really need and use a substantial amount of information? Should we still think that more wealth and

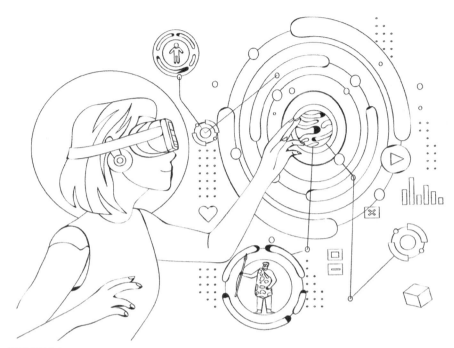

FIGURE 8.1
Human choices in the past and in the modern age.

prosperity will keep mankind continuously marching toward a higher level of advancement? These are some essential questions for today's economies that must function amid novel changes brought about by technological progress and by the new structure of human societies integrated by the internet.

Classical economic models generally assume that humans are rational, like algorithm-based robots, and will take actions that are beneficial to themselves instead of actions that are harmful or neutral (Tversky and Kahneman, 1991). In other words, it is implied in traditional economics that humans make logical and consistent choices that will optimize outcomes. It is Richard Thaler's "simple insights" that first challenged the fundamental concept that humans behave rationally when making purchasing and financial decisions, and it won him the Nobel Prize in Economics in 2017 (see Thaler and Sunstein, 2009). He is considered the father of behavioral economics, a field that disagrees with most of the classical economic models with respect to the underlying assumption of rationally optimization-seeking individuals. The sociopolitical and technological environment is changing fast for humans, and so should human behavior in response. In this chapter, some of the areas where humans are changing or not changing, contrary to intuition, will be overviewed in the light of scientific studies in behavioral economics, neuroscience, and other related fields that study human choices.

First, in the next two sections, two major assumptions of the rational choice theory will be addressed, namely the role of information (i.e., the assumption of perfect information) and the role of money (i.e., the assumption of the more, the better). While in theory these two factors directly influence decision-making, in reality humans often make choices contrary to these standard beliefs or assumptions. Although the information is vital in rational decision-making, economic actors do not always fully utilize all the information they have access to. Second, although money is a major reward for economic activities, there are plenty of other areas of decision-making where money does not motivate people or even sometimes disincentivizes them.

HUMANS DO NOT LIKE MORE INFORMATION

Technology has made information cheaper and more readily available than at any other time in human history; one example is the modern-day corporate success in becoming better, faster, and cheaper than those in the past, when information was relatively scarce (Ismail, 2014). Taking advantage of the information age, businesses in today's world make more educated decisions. Not only businesses, individuals also enjoy more access to information than ever before. This means that humans of today's generation should simply become more rational because rational behaviors rely on the availability of information and its optimal usage (Harsanyi, 1980).

At the same time, today's humans are also overloaded with information and may even suffer from something often dubbed "information or communication overload" (Lee et al., 2016). This information and communication fatigue, often caused by the widespread use of social media, may not be very conducive to better decision-making, for two main reasons. First, as mentioned before, people may be stressed from too much information and communication and may no longer seek any other available information that could lead to more accuracy. Second, too much information also means much of this information is unverifiable and unreliable; the so-called rise of "fake news," while information literacy (i.e., understanding the role and power of information), is much more important in today's world (see Behrens, 1994; Shu et al., 2017).

Let us take one example that lies between politics and business to illustrate this information problem. In much political rhetoric, Apple is often blamed for moving its manufacturing facilities from the U.S. to China, often citing corporate greed, cheap labor cost, worker exploitation in developing countries, etc. Apple CEO Tim Cook attempted to clarify in interviews that it is not because of cheaper costs but that China ceased to be a cheaper cost source of production many years prior. It is due to the availability of a huge number of tooling engineers in one place that provides China the greatest advantage, which the U.S. lacks. Even if the CEO is not entirely correct, hardly any American or any politician is likely to search for more information to verify Cook's claim and understand Apple's bias toward production in China. People will most likely be happy with the political rhetoric and will deliberately avoid looking for sufficient information and utilizing it to make an educated take on this matter. This mental bias of not using information is convenient and this inherently violates one of the major conditions underlying rational choice—perfect information. Although information is abundant and cheap today, people are simply not that interested in it.

MONEY CAN BE A DISINCENTIVE

The internet has made many things available free of cost, which discourages many content producers, such as free blog posts or other online content created by users themselves. However, the present era has turned a larger part of the population into creators than at any other time in human history. If someone has a simple smartphone, s/he can contribute with their content on the internet, often without pay. People may not like to get paid sometimes. This may seem outlandish, and this certainly violates another underlying assumption in classical economics—that more is better, especially for goods as opposed to bads[1]. A thought experiment can be done by anyone to understand this notion of monetary disincentive. For example, many anonymous contributors write for free on the internet and are not paid at all in any form. If they were now paid $10 per blog entry, many quality writers would perhaps think their time is worth more than $10 and would discontinue writing to avoid appearing as cheap contributors, although they are currently contributing their high-quality work for free. In this instance, $0 is preferred over $10,which defies the underlying assumption of rational behavior that more money is better.

In the study by Gneezy and Rustichini (2000), the authors showed that when parents of a child daycare were told about a newly introduced fine for late pickup of their children from school, the occurrences of lateness actually increased, because the penalty had now legitimized the lateness and provided parents with the impression that they were paying for the lateness. The previous guilt that they had for late pickups without a fine had now disappeared. Hence the fine could not be a sufficient incentive for them to be timely; rather, it worked in the opposite direction. Money worked in the opposite and unintended direction in this case.

In the following two sections, two crucial changes in modern societies will be addressed in light of the decision-making process. The first is the lack of motivation and the lack of sufficient creative endeavors in the digital age, when many of our activities are automated or are taken over by machines, for example. This ultimately takes away much of the motivation from many economic activities. The second one is about the fact that humans are stressed chronically in the modern age, which affects their decision-making process, a phenomenon that is probably quite different from what humans have experienced in the past.

MOTIVATION IS BECOMING RARE IN THE DIGITAL AGE

Division of labor is much more paramount in today's world than ever before. Jobs in today's world often do not produce any complete output, which would provide workers with the pleasure of creation, not just pay (Harari, 2015). Some engineers in California design a new prototype or often just a fraction of the prototype of a complex electronic device, some chip makers in Taiwan produce the chips needed, and some wage earners near Beijing assemble the final product. No one sees the whole process from the beginning to the end in this complex production process, which essentially deprives each of its makers of the pleasure of creation. When we think of happiness, we often think of a quick, pleasurable experience, such as watching a movie, drinking, etc., which overlooks the fact that these pleasures do not always produce motivation. Motivation is instead a very complex process that is often derived from a complex set of activities and sometimes even from painstaking dedication (Ariely, 2016). In today's fast-paced world, monetary payments are mistakenly thought to be the primary goal.

However, Ariely (2016) conducted a behavioral experiment in the lab, asking people to build Lego Bionicles, and showed that monetary payments often are insufficient to motivate people to work. The experiment divided Lego builders into two groups. The first group was asked to build the first Bionicle for a cash outlay, then another one for slightly lower pay, and continued to do so until the pay became sufficiently low and participants gave up. Then the second group was similarly asked to build the first Bionicle for the same first cash outlay, which was reduced similarly in each of the subsequent rounds. The only difference for the second group was that while they started building the second Bionicle, they saw their first Bionicle being dissembled by the experimenters. Then in the third round, these participants were asked to simply reassemble their now disassembled first Bionicle and the game continued to repeat using only these two Bionicles. The first group and the second group were doing the same task with the same pay level; still, the second group was found to become demotivated quickly and gave up on the game relatively earlier. The same cash outlays were not enough to keep the second group going, since it did not find any meaning in its work.

Ariely (2016) argued that motivation is a complex mix of many factors, and money only cannot explain it to any substantial degree. Other factors such as meaning, a sense of achievement, progress, the welfare of others, etc., are needed to create motivation. This is particularly important in today's society, as money is often emphasized more than meaning or purpose, and social prestige is more sought after than a sense of achievement or welfare. Technological progress has also taken away direct human involvement in many of today's real-life economic activities. Driverless cars could surely be safer and more economical than cars driven by a human; however, they probably cannot provide the warmth of care that children feel when a caring parent is driving them home.

ACUTE VERSUS CHRONIC STRESS AND RISK AVERSION BEHAVIOR

Schweda et al. (2020) ran a lab experiment involving "healthy young men" to study the effect of acute stress on their generosity in an "endowment allocation" game. In the case of the "take frame," where participants were asked to take away money from strangers, it was found that acute stress reduces generosity in individuals compared to their non-stressed counterparts (control group).

Today's professional setting is immensely efficient; humans are stressed as well, probably to the same degree. Modern-day professionals enjoy a high-flying lifestyle but, at the same time, they suffer from numerous chronic stresses. Given the existence of these chronic types of stress, decision-making could also be flawed. For instance, Veszteg (2021) found in a laboratory experiment on human subjects that men's aversion to risk reduces when they face acute stress, although women do not show any substantial behavior change. This indicates that perhaps the stressful setting of today's workplace is conducive to many risky behaviors, including bad investment and relationship decisions, etc.

In the following two sections, two developing ideas that can solve some of the humans' motivational and stress-related problems are discussed. These two factors have been found promising through careful investigations by researchers so far, and more research is also rightly underway.

WORK FROM HOME AND PRODUCTIVITY – A PANDEMIC-INDUCED REALIZATION

The pandemic of 2020 has made humans realize faster why working from home (WFH) or anywhere (WFA), for that matter, is the future of a knowledge-based economy. In addition to tangible benefits, such as saving on real estate costs, solving immigration issues, finding global talents, etc. (see Choudhury, 2021), WFA perhaps comes with a lot of intangible benefits as well. An experiment in China by Bloom et al. (2015) showed that when employees were allowed to WFH, their productivity rose by 13%. Later when they were again given the choices between WFH and returning to the office, those who chose to stay home were found to be 22% more productive. Similar findings have been found in other influential studies as well in recent years.

The question remains then: Why does WFH lead to higher productivity, especially in knowledge-based organizations? Productivity is generally viewed as an outcome of higher human capital; that is, education, advanced training, higher monetary incentives, etc. However, there are more human sides to it than just dry factors. As discussed earlier in the Lego-building lab experiment, monetary incentives cannot always keep people motivated. Choudhury (2021) identified some interesting factors that play important roles in productivity enhancement in the case of WFA. Many of the author's respondents mentioned soft areas, such as the quality of life, geographic flexibility, ability to travel freely, enabling children to regularly visit grandparents, being close to family, etc., which improve overall happiness and may influence productivity positively to a substantial degree. According to some respondents, WFA allows people to live in their favorite part of the country, which creates a great work-life balance since they get more opportunities to "relax."

MEDITATION AS THE NEW TOOL

The new generation of workers will probably be required to know about their mind and body better as well; as we progress more toward an increasingly knowledge-based economy. Gamers, for example, are generally notorious for their withdrawn lifestyle, according to their parents at least. However, gamers have been shown to gain improved eye-body coordination, enhanced reflex action, and agile thinking (see Adachi & Willoughby, 2017; Granic et al., 2014). Gaming, in a way, might train the new generation for new-age challenges that require faster decision-making compared to the previous generations. Gamers of today are probably better prepared in some sense than non-gamers to tackle the even faster lifestyle that is waiting for humans in the coming decades and centuries. If gaming can help the youth develop skills to deal with the fast-paced world, how then can humans wind down, take a step back and reflect—in other words, how can they declutter their minds from all the hustles and bustles of today's world?

Meditation and other similar mindfulness training may have the answers. Meditation helps declutter the mind and enhance concentration. Zeidan et al. (2010) conducted a lab experiment on participants with three days of 20 minutes per day of meditation intervention and compared their pain rating and sensitivity to pain (electrical stimulation). The authors also compared these results with results from participants with other interventions, namely math distraction and relaxation. Their study found meditation and math distraction to significantly reduce pain rating and sensitivity to pain. The authors concluded that mindfulness and reduced anxiety were the factors signaling meditation's analgesic effects, which in turn enhanced subjects' ability to focus on the present moment.

Seppälä (2017) highlighted the growing interest in meditation among CEOs and other senior executives in a Harvard Business Review article, which also cited a seminal paper by Miller et al. (1995) that has clinical proof of the statistically significant impact of mindfulness meditation intervention on their outpatients' subjective and objective symptoms of anxiety and panic. At the personal level, a growing number of ordinary individuals, not just CEOs, can nowadays easily download many apps such as Calm, Insight Timer, Headspace, Buddhify, Unplug, etc. The digital age has provided us with a lot of stress-related issues; hence it is no wonder that many solutions are also digital these days. Guided meditations, such as breathing techniques and guided light yoga, etc., have been found clinically and significantly effective in many scientific studies (see Benor, 2006; Levine, 2010). According to the Himalayan Yoga Institute, Bhastrika or Bellows Breath is a simple guided breathing technique that can be mastered by anyone but under some supervision, which may "help draw prana (the life force) into the body and mind, thus clearing out mental, emotional and physical blocks"(Himalayan Yoga Institute, n.d.).

This is just an overview of what many thought leaders, including CEOs, are doing to deal with the new work stresses and other mental and physical obstacles. However, it can certainly be said with conviction that humans should prepare themselves better for a work-life balance at this critical point of the history of mankind, and prepare their minds and body for a fulfilling overall existence.

SUMMARY

This chapter illustrates two significant caveats of the theory of rational choice. In the modern age, although information is abundant, humans are overburdened with information and often do not utilize all the information they have. Money has also been shown not to act as a motivator all the time. Instead, motivation is a very complex area within the human mind. Especially in today's world, where the work environment is immensely complex, motivation is increasingly becoming a rare commodity. Conscious efforts and awareness are probably needed. Without motivation and under stress, economic decision-making could be flawed. This concern may require modern-day professionals to adapt to new ways of making economic choices.

This problem also opens up various opportunities, such as the revolutionary concept of WFH, particularly induced by the 2020 pandemic. Another great example of the recent change is the growing interest in meditation, personal wellness, etc., among individuals. These phenomena indicate an urgent need for more research work to particularly look at human decision-making in the complex modern-day work environment. Knowing about own one's motivational neuropathy, and a better understanding of the body and mind might go a long way in making today's humans more resilient, free, and confident—even for making utterly dry economic decisions, such as finances, allocation of resources, and risk-taking.

NOTE

1 Bads are opposite of goods, meaning these are consumptions that provide dissatisfaction instead of satisfaction.

REFERENCES

Adachi, P. J., & Willoughby, T. (2017). The link between playing video games and positive youth outcomes. *Child Development Perspectives*, 11(3), 202–206.
Ariely, D. (2016). *Payoff: The Hidden Logic that Shapes Our Motivations* (First TED Books hardcover edition). TED Books/Simon & Schuster.
Behrens, S. J. (1994). A conceptual analysis and historical overview of information literacy.
Benor, D. J. (2006). *Personal Spirituality: Science, Spirit and the Eternal Soul, Vol. 3*. Wholistic Healing Publications.
Bloom, N., Liang, J., Roberts, J., & Ying, Z. J. (2015). Does working from home work? Evidence from a Chinese experiment. *The Quarterly Journal of Economics*, 130(1), 165–218.
Choudhury, P. R. (2021). Our work-from-anywhere future. *Defense AR Journal*, 28(3), 350–350.
Gneezy, U., & Rustichini, A. (2000). A fine is a price. *The Journal of Legal Studies*, 29(1), 1–17.
Granic, I., Lobel, A., & Engels, R. C. (2014). The benefits of playing video games. *American Psychologist*, 69(1), 66.
Harari, Y. N. (2015). *Sapiens: A Brief History of Humankind*. New York: Harper.
Harsanyi, J. C. (1980). Advances in understanding rational behavior. In *Essays on Ethics, Social Behavior, and Scientific Explanation* (pp. 89–117). Dordrecht: Springer.
Himalayan Yoga Institute. (n.d.). *9 Yogic Breathing Practices for Mind-Body Balance and Healing*. Retrieved April 25, 2022, www.himalayanyogainstitute.com/9-yogic-breathing-practices-mind-body-balance-healing/
Ismail, S. (2014). *Exponential Organizations: Why New Organizations Are Ten Times Better, Faster, and Cheaper than Yours (and What to Do About It)*. Diversion Books.

Lee, A. R., Son, S. M., & Kim, K. K. (2016). Information and communication technology overload and social networking service fatigue: A stress perspective. *Computers in Human Behavior*, 55, 51–61.

Levine, S. (2010). *Guided Meditations, Explorations and Healings*. Anchor.

Miller, J. J., Fletcher, K., & Kabat-Zinn, J. (1995). Three-year follow-up and clinical implications of a mindfulness meditation-based stress reduction intervention in the treatment of anxiety disorders. *General Hospital Psychiatry*, 17(3), 192–200.

Schweda, A., Margittai, Z., & Kalenscher, T. (2020). Acute stress counteracts framing-induced generosity boosts in social discounting in young healthy men. *Psychoneuroendocrinology*, 121, 104860.

Seppälä, E. (2017). How meditation benefits CEOs. *Harvard Business Review*. https://hbr. org/ 2015/12/how-meditation-benefits-ceos. Accessed, March 2022.

Shu, K., Sliva, A., Wang, S., Tang, J., & Liu, H. (2017). Fake news detection on social media: A data mining perspective. *ACM SIGKDD Explorations Newsletter*, 19(1), 22–36.

Thaler, R. H., & Sunstein, C. R. (2009). *Nudge: Improving Decisions About Health, Wealth, and Happiness*. London, UK: Penguin, https://doi.org/10.1002/pa.2075

Tversky, A., & Kahneman, D. (1991). Loss aversion in riskless choice: A reference-dependent model. *The Quarterly Journal of Economics*, 106(4), 1039–1061.

Veszteg, R. F. (2021). Acute stress does not affect economic behavior in the experimental laboratory. *PloS One*, 16(1), e0244881.

Zeidan, F., Gordon, N. S., Merchant, J., & Goolkasian, P. (2010). The effects of brief mindfulness meditation training on experimentally induced pain. *The Journal of Pain*, 11(3), 199–209.

9

New Work and Hybrid Forms of Collaboration – Corporation or Not?

Marc Nathmann
Legal counsel and attorney at law, ING Germany, Frankfurt, Germany

CONTENTS

NEW WAY OF WORKING – ANYWHERE YOU LIKE

New technologies often contain the promise of "work from anywhere you like." In fact, it is currently technically possible to work virtually not only outside the office but actually in teams that are distributed all over the world, not only in Europe, due to the increased spread of home offices as a result

FIGURE 9.1
Mobile cross-border collaboration.

of Covid. In a virtual meeting, it is, in fact, irrelevant where the participants are physically located. In addition, virtually all widely used IT systems and programs necessary for work provide easy access from almost anywhere in the world. This chapter, for example, was created in the USA, Portugal, Germany, and Netherlands on different devices.

Increasingly, collaboration is also taking place in purely or partially virtual teams. Collaboration in virtual teams does not only take place within a fixed corporate structure (i.e., within an employment relationship); rather, it is increasingly the case in projects that collaborators come together outside of contractual relationships agreed upon and regulated under employment law or corporate law.

Practical skills can quickly come into conflict with legal permissibility. A wide variety of legal issues arise in connection with cross-border cooperation in particular. Most significant in terms of legal implications are the rights and obligations of persons that may arise from cooperation without a fixed legal structure and in an international context. This article, therefore, aims to identify the main legal hurdles and point out possible solutions.

GENERAL CHALLENGES WITH VIRTUAL WORKING – INFINITE WORKING?

For employees in particular, working in virtual teams can mean enormous flexibility; at least in theory, work can be scheduled around free time. Furthermore, it is often a dream idea to work from the most beautiful places on earth. Worldwide flexible working sometimes also means working in different time zones, which has a considerable impact on working hours. As a result, flexibility also quickly leads to a lack of a fixed framework of fixed working hours. Flexibility also means saying goodbye to nine-to-five working system.

The well-known quotation from Milton Friedman "there is no free lunch," might be true for virtual working as well. The tremendous promise of "work from where you like" and even more "when you like" might be paid with (probably unpaid) overtime hours. Since especially WFH has become, at least during the Covid lockdown, like a new normal for employees, legal literature on the boundaries of WFH is quite rare, as is the legislation on such. Most labor laws simply regulate working times without taking into consideration that work and spare time are increasingly merging into one another (see Katsabian, 2021, p. 418).

In the case that employees are offered a free configurable virtual working environment within the framework of collaborations, the question arises as to how, in particular, labor law requirements can be reconciled with maximum flexibility. In no relevant country is there a regulatory playing field on a flexible and freely configurable working environment, even though the new virtual reality exists. However, not only as a proposal for lawmakers, a high level of flexibility and independence of both employer and employee can be achieved by trustful cooperation on each side. Crucial is the count of every actual working time unit, including tracking of overtime. In particular, the value of overtime should derive from negotiations between the employer, the employee, and, given there are such, the employee representatives, and it can be adjusted to reflect the needs of specific positions and workplaces. In addition, the right to disconnect is essential to avoid ambiguity and enables more flexibility and autonomy for both the employee and the employer through its default sections (see Katsabian, 2021, p. 418).

The bottom line is that virtual work requires mutual trust and fairness between employer and employee. In the absence of explicit specific regulations under labor law, the parties involved should reach a common understanding of the framework conditions of work, particularly with regard to time. One idea may be to agree on goals rather than rigid time conditions.

The above is basically significant from the perspective of the employer-employee relationship. Nevertheless, even in forms of free collaboration between the participants, clear rules should be made about time cooperation. Precisely because self-employed persons often lack rules under employment law, a contractual framework is required for the framework conditions of collaboration in terms of time. Analogous to the proposal in the employment relationship, the collaborators should make clear regulations about times of availability and non-availability and set clear goals for each individual collaborator. This applies regardless of the form of collaboration under company law discussed in the following.

_____

LEGAL FORMS OF COOPERATION

Free Collaboration

An essential question is whether and to what extent the collaboration creates reciprocal rights and obligations between the collaborators; that is, what legal form the collaboration takes.

As a first step, all parties involved should consider whether they wish to be mutually bound. If no such commitment is to be made, only the form of free cooperation is suitable.

In the case of free collaboration, the rights and obligations between the collaborators are freely designed and agreed upon without a legal form. Rights and obligations then arise exclusively in the internal relationship between the participants but not in the external relationship vis-à-vis third parties. This means that each of the collaborators is basically liable for her/himself.

As a rule, free collaboration must be extensively regulated individually by contract. The collaborators are legally opposed to each other as natural persons. In particular, questions of the common goal and purpose of the collaboration and who is entitled to which rights must be agreed upon individually between the collaborators. The distribution of profits/income and the rights to intellectual property (IP) will cause particular difficulties. Individual case regulations are necessary here.

Likewise, the conclusion of contracts with customers and clients will create hurdles in a way. The collaborators must decide and ensure which of them is to become a contractual partner.

It can be roughly stated that, in the case of free cooperation, the entire mechanism of cooperation must be contractually regulated on an individual basis.

Corporation

The other variant of collaboration is a corporation. This enables the collaborators to bind each other in a legally fixed structure. In their external relations with third parties, the collaborators are therefore mutually liable for each other, insofar as and to the extent that they are liable for their corporation. The corporation provides the legal bracket and the mechanism of cooperation. As a rule, with the exception of the respective national law, companies can legally bind themselves independently and, in particular, can also be contractual partners. For collaborators, in simplified terms, their corporation acts as a representative of the outside world.

The corporation enters into contractual relationships with customers and clients. Profits may be retained. The corporation may also be entitled to IP rights.

National corporate laws generally establish a clear framework for the rights and obligations of collaborators as shareholders.

Demarcation Between Free Cooperation and a Corporation

In legal theory, it seems trivial that free collaboration and corporation simply depend on the will of the collaborators. A corporation is created by the collaborators corporating their collaboration by concluding articles of association.

More risky is the question of when intended free cooperation does not result in a corporation. German law, for example, also recognizes implied contracts. It is possible to conclude a partnership agreement implicitly, even through occasional cooperation. This is explained by the character and content of these agreements. The decisive criterion for the question as to whether, in these cases, the formation of a corporation is to be assumed on the basis of the actual cooperation of the parties or whether the cooperation takes place on an extra-legal, purely voluntary basis is primarily the economic interest of the parties involved. Accordingly, a corporation can come into being "by accident" if the collaborators work closely together and pursue a common goal.

If, therefore, a corporation is not to be founded, the collaborators must ensure that, in case of doubt, they pursue their own economic interests in addition to their common interests. The essential legal indication here is that each individual collaborator first pursues their own economic interests beyond the common purpose.

INNOVATIVE CORPORATE FORMS – DECENTRALIZED AUTONOMOUS ORGANIZATION

In addition to traditional companies, corporations, or forms of collaboration, new types of decentralized autonomous organizational forms based on self-executing smart contracts (so-called "DAOs") are slowly emerging. The first successful example of a DAO was BitShares, a virtual e-commerce platform linking merchants and customers without a central authority.

Such companies run without people; they do not have a CEO, a board of directors, managers, or other decision-makers. No people at all are involved. The group votes to make decisions, but this process happens automatically. It is an approach to control corporations via the "crowd" of investors in a grassroots democratic way. DAOs come in many structures, but all operate as collectives in which members make decisions democratically. No single person exerts control in the way a conventional CEO or senior management team would (Ruane & McAfee, 2022).

Based on smart contracts, complex entities can be created from interlinked blockchain-based smart contracts. Smart contracts are computer protocols that can map or verify contracts or technically support the negotiation or settlement of a contract. Smart contracts usually also have a user interface and technically map the logic of contractual regulations (Schüffel et al., 2019).

This makes it possible for people to join together to form a corporation whose processes and rules are controlled in an automated manner by a previously programmed computer code and stored in a decentralized manner on the blockchain without the need for centralized management. The management of the board transaction is executed in an automated manner using smart contracts.

A DAO is tailor-made for alternative financing options, namely crowdfunding, due to its decentralized grassroots democratic and open "corporate" and decision-making structure.

Interested investors can even provide this DAO with capital in the form of cryptocurrencies. The organizational form takes advantage of the blockchain. All data is stored immutably in the blockchain, and no trusted third party is needed to hold the asset amounts, and necessary funds can be kept available by freezing. The potential of smart contracts to reduce transaction costs can then manifest itself within a DAO as the potential to mitigate intra-organizational transaction costs (Aufderheide, 2022).

The exact legal status of this type of business organization is unclear. Some similar approaches have been deemed illegal offerings of unregistered securities by the U.S. Security and Exchange Commission. Although vague, a DAO may functionally be a corporation without legal status; that is, a general partnership. This implies potentially unlimited liability for participants (Chohan, 2017). However, there are some very slight approaches existing to put DAOs on a legal basis within the U.S. company law system. In July 2021, Wyoming became the first state in the U.S. to explicitly codify rules around DAOs wishing to become domiciled in that jurisdiction. This rule change means that DAOs in Wyoming are considered a distinct form of limited liability company, which grants them a legal personality and confers a wide range of rights, such as limited liability for members (Ruane & McAfee, 2022). The push from Wyoming certainly seems like an interesting approach to fitting DAOs into the legal system. Nevertheless, in the U.S. and even more so worldwide, this is a first and isolated approach to quasi-legalizing DAOs.

While it is conceivable that DAOs can be classified under national company law in individual cases, as a rule it will be legally nothing or at least a legal gray area. This makes it not only difficult but also extremely critical in terms of liability law as soon as contractual relationships are entered into with third parties. It is simply completely unclear whether a DAO can legally bind itself at all.

Against this background, DAOs still appear to be more of an idea than a form of collaboration. Due to the relatively unclear legal situation, classic fundraising at a DAO will be rather doubtful. Although a DAO is ideal for alternative (crowd) financing forms, alternative fundraising via the issuance of cryptocurrencies will also be the only capital acquisition option, at least as of today.

ORGANIZING COLLABORATION BY USING SMART CONTRACTS

Even if DAOs in their pure form may still appear to be a vision of the future, one of the technological foundations, smart contracts, can certainly be helpful in increasing the efficiency of virtual collaboration. First, smart contracts can result in reduced transaction and legal costs. The absence of any central authority or trusted intermediary in a blockchain means that many of the numerous transactions and legal costs that would normally be incurred through intermediated transactions are removed. Such fees are typically in the nature of service or administration fees or legal costs associated with the preparation, supervision, and execution of written contracts. In addition, smart contracts enable a high degree of anonymity and transparency for all parties involved. (Giancaspro, 2017).

Smart contracts are highly suitable for the (automated) processing of service relationships. Smart contracts can be used to automatically trigger consideration when a specific service is performed. For example, the fulfillment of an element of a contract can automatically trigger the payment. Smart contracts can also be used for automated contracting in the case of recurring services or long-term contracts. One of the main application areas is to map service-level agreements. Some peer-to-peer networks require mechanisms to ensure that remote partners contribute as much as they consume without the overhead of written contracts.

Collaboration in particular opens up a wide range of application areas; for example, in the measurement of performance-based remuneration elements that determine the amount of remuneration depending on individual performance. A smart contract could thus trigger the payment of the corresponding remuneration per unit of work performed (e.g., the factors quantity, distance, and savings can be measured). One of the main advantages in practical terms is that documentation of work gives results; payment and documentation are automatically stored and made transparent to all collaborators.

Since the blockchain on which the smart contract is based would record the point in time at which the number of parts still available is so low that replenishment is required, it would be possible to draw conclusions about the pace of work. This would make it possible to measure the efficiency and the share of the respective collaborator on the overall result and the share of work on the joint project. Particularly if the collaboration is to be organized as free collaboration, smart contracts can not only regulate essential areas of collaboration but also control and execute them efficiently with minimal effort.

The legal classification of smart contracts is always a question of the individual case and naturally depends on the respective national law.

Internationally, however, smart contracts are not contracts in the legal sense. In particular, a clean subsumption under the general rules is necessary for the legal treatment of smart contracts.

A contract is a legal construct based on offer and acceptance, the content of which is to be determined by way of an overall view of all the circumstances of the individual case and in accordance with the principles of good faith. A smart contract is merely a computer program that represents a contract (Raskin, n.d.). The code contained in the computer program may be the language of the contract, the equivalent of a contract document, and an instrument of contract execution. In general, the following construction of the contract seems legally conceivable: The legal conclusion of the contract is preceded by the use of the smart contract, so that the smart contract merely becomes a processing modality that controls, monitors, and documents the exchange of services. In this respect, a comparison with a vending machine that executes the agreed exchange of services is appropriate. Smart contracts are thus regularly embedded in legal contracts, which suggests a description as a performance mechanism.

PROS AND CONS OF CHOOSING THE FORM OF COLLABORATION

Which form of collaboration the collaborators choose is always an individual decision in each case. However, the following general parameters can give guidance:

Pro corporation:

- Raising capital is generally easier due to the possibility of issuing shares and greater confidence on the side of lenders and investors.
- As a rule, comprehensive limitation of liability is only possible as a corporation.
- Legal relationships with third parties can be concluded centrally via the corporation.
- Firmly structured legal relationships among collaborators through national corporate law and bylaws.

Pro free collaboration:

- Free cooperation is to be legally more flexible and individualized.
- There are no minimum capital requirements.
- Usually lower requirements for registered entries and lower formation costs.

As a "rule of thumb," the larger and more investment-intensive the project, the more likely a corporation is the form of choice.

CHALLENGES IN CROSS-BORDER COLLABORATION

In addition to the legal form, hybrid collaboration is often characterized by a cross-border element. If collaborators are located in different countries, different legal systems apply in each case. The presentation of all relevant conflict-of-law rules would go beyond the scope of this article, but the question of which legal rules each individual collaborator has to observe is relevant. Since hybrid collaboration is usually an economic one, a choice of law is usually possible. This means that the collaborators can, in principle, agree bilaterally and in relation to customers or clients which law is to apply. It is important that such a choice of law must be expressly regulated in the contract in order to be legally effective.

As far as the internal relationship is concerned, in the case of free collaboration the collaborators can contractually regulate which law they want to apply for or against themselves.

In the case of a corporation, the question is somewhat more complex. There is no uniform international company law. The law applicable to the corporation, and thus in particular to the relationship between the shareholders, is defined by the state in which the corporation is domiciled. In this context, the domicile of the corporation is of decisive importance. There are some states that apply their law to such companies that have been formed or entered into public registers in that state. In addition, there are states that base their law on the registered office, in other words the place where the main operations take place and/or where the corporation is managed (the so-called seat theory). If a certain law is to be applied to the corporation, the national company law in question must always be analyzed (Pargendler, 2021). If a state applies the domicile theory, the shareholders will have to be in that state at least regularly for essential decisions, which in practice means presence on the spot.

In the external relationship of the corporation (i.e., vis-à-vis customers and suppliers), a choice of law is generally possible without any problems, given the counterparty is deemed a commercial customer.

Which law should be applied depends on many factors and ultimately on the individual preferences of the business model and the collaborators. In particular, the practicability of the respective law for the business model plays a decisive role.

It is also important for the collaborators to include jurisdiction agreements in all contracts. In such clauses, the parties can specify whether, for example, state courts or arbitration courts (private) should decide on disputes. Particularly when the business models are more complex, arbitration courts can be a good choice due to the higher expertise of the judges. In the case of arbitration courts, however, it should be noted that proceedings heard there are generally somewhat more expensive than in state courts. Therefore, in practice, a commitment to arbitration courts will only make sense from expected subject matter values of the dispute of about U.S.$1 million. The use of an arbitration court is indispensable for countries with a problematic or weak legal infrastructure if state courts offer little guarantee of due and fair proceedings.

IMPORTANT INDIVIDUAL LEGAL CHALLENGES

Data Protection – GDPR

In addition to the above rather general considerations, data protection law poses a challenge for cross-border collaboration that should not be underestimated.

Hybrid collaboration is generally characterized by the increased use of online communication. Insofar as personal data is (also) transferred in the process, the scope of application for data protection regulations is opened up. Data protection law has become much more strictly regulated internationally in recent years, including fines and other sanctions for violations. A good example is the GDPR (Regulation (EU) 2016/679, General Data Protection Regulation) applicable within the EU. Apart from specific rules on whether and under what conditions data may generally be processed, data protection laws generally contain restrictions on the cross-border transfer of data.

Within the scope of the GDPR, it must be checked whether a transfer to a third country is permissible. A distinction is made between safe and unsafe third countries. Safe third countries are those to which the EU has confirmed an adequate level of data protection by an adequacy decision. There, national laws guarantee a level of protection of personal data that is comparable to that of EU law.

Safe third countries include: Andorra, Argentina, Canada (commercial organizations only), Faroe Islands, Guernsey, Israel, Isle of Man, Jersey, New Zealand, Switzerland, Uruguay, Japan, the United Kingdom and South Korea. Data transfer to these is therefore expressly permitted.

Outside of secure third countries, data transfer is difficult and requires an adequacy decision stating that a level of data protection comparable to the EU exists in that state.

If no adequacy decision exists for a country, this does not in principle preclude a transfer to that country. Rather, the processor must ensure by other means that the personal data are adequately protected at the recipient's site. This can be done through the use of standard data protection clauses, in the case of data transfers within a corporate group through so-called "binding corporate rules", through an obligation to comply with codes of conduct that have generally been declared applicable by the EU or through certification of the processing operation (see Auer-Reinsdorff & Conrad, 2019, § 35 II. 4. c).

IT Security

Similar problems in the area of data protection for personal data also apply more generally to IT security requirements, which include, in particular, the protection of data classified as confidential.

General duties of care in the context of IT security can be found in corporate and commercial law; in Germany, for example, the Aktiengesetz (German Stock Corporation Act). In addition, customers or clients will by default contractually impose IT security due diligence obligations. Such duties of care include, in particular, the identification and combating of IT security risks. General duties of care with regard to IT security include early risk identification and risk management, the establishment of preventive security measures, and duties to provide information in management reporting (Leupold et al., 2021, p. 737).

In concrete terms, such obligations are often structured as in the German regulation on critical infrastructures and cover the availability, integrity, authenticity, and confidentiality of the IT systems used. They often do not specify any concrete protective measures to be implemented. Instead, it is left to the collaborators to determine the specific security measures on the basis of the standardized security objectives on their own responsibility. As a rule, IT security requirements thus follow a risk-based approach comparable to that of data protection law (Voigt, 2022, Chapter D). The determination of necessary and suitable measures to achieve the security objectives necessarily requires a risk assessment to be carried out in advance (Leupold et al., 2021, p. 742).

Copyright

In any form of collaboration—hybrid or classical—in which works worthy of copyright protection are created, copyright is a central issue. Essentially, two relevant groups can be distinguished:

- Compilations: This group is characterized by the fact that, although the creative contributions of the individual authors remain clearly separable here as well, the individual contributions remain essentially unchanged. This category includes compilations and collective works under U.S. law.
- Collaborative works: In these groups, at least two natural persons work together in such a way that a joint end product is created. There is the possibility here that the individual creative contributions can no longer be clearly assigned to one person at the end (Strömholm, 2007, p. 349).

If the collaborators are to be jointly entitled to copyrights, this is generally unproblematic. However, legal classification problems arise if only individual collaborators are to be entitled to copyrights. In this case, some jurisdictions provide that, in case of doubt, the copyright is jointly owned by all co-creators (e.g., Austria).

Cross-border collaborations also cause difficulties in copyright law. In cross-border constellations, however, the international Rome II Regulation can help. Article 8 (1) of the Rome II Regulation is a new conflict-of-law rule for copyright infringements. Within the scope of the Rome II Regulation, the law of the country of protection applies. According to this, copyright protection can be enforced in the country for whose territory the protection is claimed. Accordingly, the registration of the respective right under public law is decisive.

SUMMARY

The key insight with regard to the hybrid collaboration discussed here is: The technology may change, but legally everything remains the same. In a hybrid form of collaboration, the key question is in which legal form the collaborators want to bind themselves:

- Maximum flexibility as free collaboration or
- Legally secure in a corporation

Whereas DAOs do not yet represent an alternative to traditional corporate forms, legal developments in the U.S. should be monitored. Particularly with a view to a flexible and grassroots form of company and alternative financing options, the DAO can offer itself as a genuine alternative. This perspective remains exciting in the (hopefully) very near future. Until that point, free collaboration can also be highly individualized via smart contracts. These offer the option of efficiently executing an individual framework at favorable transaction costs.

In the here and now, legally significant problems will arise not so much from the use of new technologies but rather from the crossing of borders in the collaboration. The question of the choice of law must therefore be carefully considered in the run-up to the collaboration.

One challenge arising from the use of new technologies is posed by IT law issues. Documentation and monitoring obligations play a role here in particular: Collaborators must ensure that data provided to them is secure at all times and also document this.

REFERENCES

Auer-Reinsdorff, A., & Conrad, I. (Eds.). (2019). *Handbuch IT- und Datenschutzrecht* (3. Auflage). C.H. Beck.

Aufderheide, S. (2022). Dezentrale Autonome Organisationen (DAO) – Smart Contracts aus der Perspektive des Gesellschaftsrechts [Decentralized Autonomous Organizations – Smart Contracts from the Perspective of Corporate Law]. *WM Zeitschrift Für Wirtschafts-Und Bankrecht [WM Journal of Business and Banking Law]*, 6, 264 cont.

Chohan, U. W. (2017). The decentralized autonomous organization and governance issues. *SSRN Electronic Journal*. https://doi.org/10.2139/ssrn.3082055

Giancaspro, M. (2017), Is a "smart contract" really a smart idea? Insights from a legal perspective. *Computer Law & Security Review*, 33(6), 825–835.

Katsabian, T. (2021). It's the end of working time as we know it: New challenges to the concept of working time in the digital reality. *McGill Law Journal*, 65(3), 379–419. https://doi.org/10.7202/1075597ar

Leupold, A., Wiebe, A., & Glossner, S. (Eds.). (2021). *IT-Recht: Recht, Wirtschaft und Technik der digitalen Transformation* (4., überarbeitete und erweiterte Auflage). C.H. Beck.

Pargendler, M. (2021). The rise of international corporate law. *Washington University Law Review*, 98(6), 1765–1820.

Raskin, M. *The Law and Legality of Smart Contracts* (www.ilsa.org/ILW/2018/CLE/Panel%20%2311%20-%20THE%20LAW%20AND%20LEGALITY%20OF%20SMART%20CONTRACTS%201%20Georgetown%20Law%20Technology%20Rev.._.pdf)

Ruane, J., & McAfee, A. (2022, May 10). What a DAO can—and can't—do. *Harvard Business Review*. https://hbr.org/2022/05/what-a-dao-can-and-cant-do

Schüffel, P., Groeneweg, N. & Baldegger, R. (2019). *The Crypto Encyclopedia: Coins, Tokens and Digital Assets from A to Z*. Bern: Growth Publisher.

Strömholm, S. (2007). Copyright – Comparison of Laws. In E. Ulmer & G. Schricker (Eds.), *International Encyclopedia of Comparative Law: Vol. XIV*. Mohr Siebeck.

Voigt, P. (2022). *IT-Sicherheitsrecht: Pflichten und Haftung im Unternehmen* (2. neu bearbeitete Auflage). Otto Schmidt.

10

Electronic Person – Who Is Liable for Automated Decisions?

Marc Nathmann

Legal counsel and attorney at law, ING Germany, Frankfurt, Germany

CONTENTS

TECHNOLOGIES PAIRED WITH AI

Since digitization continues to advance, decisions are increasingly being automated. Technologies paired with AI make it possible to have many decisions made fully or semi-autonomously. In general, AI can hardly be defined in a clear-cut way. AI is a mix of many different technologies. It enables machines to understand, act, and learn with human-like intelligence. For this paper, AI therefore means, in the broadest sense, all automated decisions that are largely independent of human intervention. Therefore, AI results in a transfer of decisions from humans to machines or algorithms.

DOI: 10.1201/9781003371397-10

FIGURE 10.1
Human–machine interaction.

From a legal perspective, such transfer results in the question of who is liable for AI and, above all, will AI itself be liable in the future? This chapter refers in its basic ideas to German law. However, the principal ideas shown by the particular German legal situation apply to many jurisdictions and, therefore may be, to some extent, generalized.

SIGNIFICANCE OF AI

AI is already relatively widespread. For example, in e-commerce, shoppers can be provided with additional offers in a more targeted manner. Other typical areas of application include assistance programs, language translation (e.g., DeepL), automated consulting processes, and autonomous driving. In industry, networked and automated machines (e.g., robots) or production lines are used as part of the so-called Industry 4.0. In the broader area of legal advice, so-called smart contracts solutions (i.e., computer protocols that map or check contracts or technically support the negotiation or settlement of a contract) can also be based on AI or supported by the use of AI (Kaulartz & Heckmann, 2016).

AI is considered to be of great importance in Europe and internationally. As far as industry investment in AI is concerned, the U.S. is one of the frontrunners and the EU brings up the rear internationally. In the EU region, only between €2.4 billion and €3.2 billion was invested in AI projects in 2016. This is less than half the investment in Asia and low compared to North America (especially the U.S.), where between €12.1 billion and €18.6 billion was invested in AI projects in 2016. The same is true for research on AI (Manyika, 2017). Accordingly, the AI competence of employees in European and, in particular, German companies is below average in international comparison (EY, 2019). Europe also lags behind in terms of hardware; 53.4% of all supercomputers are located in Asia. However, the absolute economic performance is significantly lower and would suggest a much higher share in North America and Europe.

Europe's lag can be explained culturally. Europe is skeptical of innovation compared to the rest of the world. In Europe, AI is perceived more as a threat, in the U.S. it is viewed as useful, while in Asia almost euphoric (Ramge & Galieva, 2022). At the same time, however, as far as the assessment of the importance of AI in the future is concerned, it can also be seen in Germany that a significant increase in importance is expected, even though only around a quarter of the companies have implemented or firmly planned AI projects. Companies expect the strongest impact in the services sector, followed by medical services and finance. As a result, almost 50% of the German companies surveyed consider AI to be an essential part of their digital strategy. The importance of AI in the respective industry is estimated even more clearly than for their own (EY, 2019).

FUNDAMENTAL LEGAL CHALLENGE – ACCOUNTABILITY

Despite the steadily increasing importance of AI, the legal situation with regard to its fields of application is still relatively undeveloped. In essence, this almost always relates to liability issues. When it comes to liability for autonomous decisions, there are three protagonists that could be considered:

- The AI itself
- The programmer/manufacturer
- The user

The first protagonist appears to be excluded. A major problem is that legal systems generally assume human responsibility. The reason is simple—legal systems in general are homocentric. They focus on and address humans. A good example is shown in the German constitution. Art. 19 para. 3 and 20a GG (German Basic Law – German constitution) clearly expresses this homocentric constitutional view. The article puts humans in the center of the environment (Steinberg, 1998). A possible own responsibility of AI would not be in accordance with such homocentric principles. The same is true for the vast majority of legal systems worldwide. Consequently, any machine decision or action in the causal chain must be traced back to a human creator (Nathmann, 2021).

In the discussion about AI, it is regularly pointed out that, in the context of AI, algorithms, programs, and machines approach human intelligence by acting in a technically autonomous way. Nevertheless, there is a lack of autonomy in the social and legal sense, which would be necessary for recognition as a protagonist. Such autonomy requires not only technical independence and autonomy but also the ability to establish one's own principles on the basis of one's own evaluations and to act according to them. The decisive difference between AI and human intelligence is that AI as a technology lacks the emotional component. Therefore, AI alone is capable of rational but not of emotional or even moral decisions (Misselhorn, 2018).

This interaction between increasingly autonomous machines on the one hand and the reliance on humans as responsible protagonists on the other hand, leads almost inevitably to the question: Can an AI or an autonomous machine be self-sufficient and autonomously liable?

CURRENT LEGISLATIVE PROJECTS

General Projects

Even though AI in the EU is not that widespread compared to other parts of the world, the slight AI skepticism in Europe results in wider regulatory efforts with regard to AI. Such efforts legally concern AI. As a first step, the EU wants to clarify liability issues and has also created a kind of definition, among others, for AI and robots with the proposal on civil law regulations in the field of robotics (*EU Commission, Draft Resolution of the European Parliament, with recommendations to the Commission on civil law regulations in the field of robotics, 2015/2103 (INL) of 27.05.2018*). AI is thereafter:

- Obtaining autonomy via sensors and/or via data exchange with their environment (interconnectivity) and the provision and analysis of these data.
- Ability to self-learn through experience and through interaction (optional criterion).
- At least a minimum physical support.
- Ability to adapt their behavior and actions to their environment.
- Not living beings in the biological sense.

On the other hand, the EU wants to address the legal challenges posed by AI and regulate the core areas of product liability, IT or system security, and data protection that it has identified. This is to be achieved through appropriate national oversight of AI users and programmers. In this context, the EU essentially recognizes that legal protection must always be guaranteed for decisions made by AI against a person responsible for the AI. Furthermore, the protection of data and information in the online area plays a decisive role.

It is clearly recognizable that the EU also wants to focus on the human protagonists behind it in its legislative plans with regard to liability for AI.

Privacy

Unmistakably linked to AI is data protection as a core challenge of digitization. In the field of data protection the EU again internationally finds its role as the frontrunner in regulation. With regard to data protection, the EU sees essential protective functions implemented by the GDPR. While the GDPR still appears quite unspecific with regard to new challenges, significantly more specific regulations are to be implemented in the ePrivacy Regulation, namely with regard to metadata that is extremely relevant for AI and explicit and clear information obligations. This is a *lex specialis* to the GDPR with regard to online services according to Art. 1 (3). The ePrivacy Regulation is intended to translate the secrecy of correspondence into the digital world and thus concretize the rules from the GDPR for the particularly sensitive area of electronic communication.

In the light of AI, the ePrivacy Regulation, like the GDPR, helps to strengthen data protection in the online area, but it cannot be applied to AI in essence. The special feature of AI is precisely that it accesses data but does not necessarily collect it itself, and AI models in particular do not regularly qualify as communication services pursuant to Art. 2 of the ePrivacy Regulation. Neither the GDPR nor the ePrivacy Regulation can solve the problems raised by AI. For these problems, the fixation on personal data is still too narrow, and there is a lack of consideration on how to deal with information.

Proposed Solution – Electronic Person

Outside the EU, regulatory as well as general legislative approaches seem more sophisticated. A completely different and quite revolutionary proposal is the idea of the electronic person. The basis of the idea is an analogy to a legal person, for example a corporation. The electronic person would then be granted its own rights and obligations on the basis of a legally arbitrary definition.

In the case of an electronic person, the liability would lie with the AI itself. AI would thus become a legal person of its own, similar to a company, with corresponding rights and obligations (Matthias, 2004).

Sometimes this demand even slides into the absurd; Bill Gates is said to have even proposed taxes for "robots." The idea is supported by gaps in civil liability recognized by the EU Commission. As a bearer of rights and obligations, the electronic person would have equal rights to the natural and legal person (Kingston, 2016). As a consequence, there would be no need for complex attributions of AI decisions to the protagonists behind them, such as operators or programmers. The AI itself would be liable for its decisions.

However, the essential basic principle for the figure of the electronic person is that the (legal) situation of legal persons (e.g., corporation) and AI is generally comparable.

CRITICISM OF THE LEGAL FIGURE

However, considerations of the electronic person fail to recognize that AI is an instrument. AI is not similar to legal persons and also does not indicate a comparable interest situation. The fundamental consideration of a legal person is that it appears as an organism in order to guarantee the protection of fundamental rights derived from individuals (Isensee & Kirchhof, 2003, Vol. V, § 108 Rn. 31).

A legal figure similar to a legal person thus presupposes that AI could be granted derived protection of fundamental rights. But this is lacking. Law in general is tailored to human beings. The person behind AI naturally enjoys the protection of fundamental rights. However, this protection is based on individual human interests. No interests are shifted to AI; it is only an instrument to make decisions in certain areas. From the user's point of view, the AI's decisions serve to implement his or her interests, which may be protected by fundamental rights. AI is thus more of an organ of the individual or collective interests. However, the interest itself remains with the protagonists behind the AI. There is no shift of interests. Even more remote would be the attribution of vested interests to AI. According to general understanding and the state-of-the-art, AI lacks intrinsic motivation. It does not fulfill its purpose out of its own motivation. AI acts according to the laws of the programmed code.

Essential to the debate about the electronic person is to point out that even strong AI, in cases where it approaches human cognitive abilities, does not allow a new protagonist to emerge (Misselhorn, 2018). AI is a purely technical implementation tool and emerges through a human act, a programming process. The programmer creates the AI. This act is the first causal step and the basis for all subsequent decisions that the AI makes autonomously. The autonomy of decisions also arises only because that autonomy is provided by the act of programming. Even strong AI cannot consciously extend decision-making authority. Rather, the decision autonomy of AI is conditioned by human delegation and is equally limited. The limitation, namely by defining the areas of use and application, is also made by another human protagonist—the user. Any sequence of AI decisions can be relatively easy to trace back in the causal chain to the user, sometimes the programmer. The user always enables the AI decision and thus its direct and indirect consequences.

Therefore, if one draws the comparison to the legal person, AI would be more comparable to the body of a legal person. One could even argue that AI concerns the opposite case of a legal person. In a legal person, individual interests are bundled; for example, in an association or a company. AI, however, is used to implement interests, in other words to make decisions.

Consequently, AI cannot be a bearer of rights and obligations. It is crucial to ensure that accountability for decisions made by AI can be enforced against the protagonists behind them.

RIGHT TO HUMAN CONTROL

Responsibility ultimately means accepting the consequences of decisions between "right" and "wrong." The special characteristic of human decisions is always also a valuation of manifold, sometimes irrational parameters due to ethical, traditional, or also cultural influences. This is particularly evident in freedom of opinion. Freedom of opinion is not based on purely rational or true statements, but protects the part that cannot be proven, ultimately individual evaluations and judgments. Precisely because, according to the basic idea, any constitution protects the individual freedom of such value decisions, but also presupposes it in its conception of man, a responsible protagonist can always only be a human being, since only he/she is capable of these individual value decisions. AI cannot make value decisions.

In the case of automated decisions that have a legal effect, there is a right to human control. The constitution mandates that a human or legal person can be held responsible for any possible decision of AI or its effect (Armour & Eidenmüller, 2020; Lauwaert, 2021). While a delegation of responsibility may change the duty situation and reduce duties to monitoring, as, for example, §§ 278 or 831 BGB show, it does not mean that an act of delegation leads to a total release of responsibility. The European Court of Justice has ruled that technical processing that is as autonomous as possible does not absolve humans from ultimate responsibility. Legal responsibility cannot be shifted to machines or technology (Nathmann, 2021).

ECONOMIC CRITIQUE

Economically, too, there is little to be said for assuming an electronic person. The function of AI as protagonist in its own right is already lacking. However, it seems fundamentally sensible to counter AI with regulatory transparency requirements in order to contribute to the elimination of information asymmetries. In addition, the formation of a data oligopoly must be countered.

In practice, however, this insight is difficult to implement. Mere, presumably difficult to present information about AI is not sufficient. Similar to other economic sectors, such as investor protection, it is apparent that information obligations are often implemented in extensive notices or prospectuses without providing relevant added value for users. The decisive protagonist is the elimination of information asymmetry with regard to consequences. The user must be made transparent about the consequences of the decision by AI and, in particular, about further consequences. This means, on one hand, the fact that AI makes a decision and, on the other hand, what the decision is based on, namely, whether the user discloses information or data for further purposes. Ideally, this would lead to self-regulation of "good" AI systems by means of negative selection and, at the same time, to an estimation of the monetary value of disclosed information (i.e., to the formation of a market and a recognizable market price for information from the user's point of view). This transparency and a market price could at the same time counteract excessive concentration if data is no longer collected "for free." There is also a close economic connection between information collection and AI.

The economic questions, however, as well as the legal ones, result from the interaction between AI and humans. Transparency should also be created with regard to this circumstance of the greatest possible rationality.

At the same time, it may be necessary to make more data freely available. This would prevent oligopolies. This does not necessarily have to be done by forcing companies to share data but by providing incentives or opening up public data sources (Nathmann, 2021).

A non-regulatory measure to increase transparency in the long term is to improve the education of users and increase their knowledge of the functions and consequences of AI. Here, parental education, school curricula, and the "learning by doing" of the following generations should be relied upon. The knowledge of broad sections of the population should keep pace with technological development to some extent.

BASIC CONSIDERATIONS ON ACCOUNTABILITY FOR AI

The starting point is to prevent protagonists from evading responsibility by transferring decisions to AI. At the same time, the principle of ultimate human responsibility must apply. A principle must be established such that the legal standards for "fair" and "right" must be incorporated into the act of programming so that these become part of the decision-making standard of AI. AI must not change the basic legal order and the essential legal standards; rather, and this is the legislative mandate, it must be ensured that these values are taken into account by AI decisions. This can be done in the framework of a voluntary commitment (Nathmann, 2021).

To this end, it is first necessary that there is always an ultimate human responsibility. This requires that a responsible addressee is defined. This can be either the programmer, the user, or both. This would follow the idea that liability must be assumed for the risk of AI, similar to the principles of product liability. Risk liability seems legitimate from the point of view that the user delegates their own decision and thus consciously releases it from their sphere of control. If the addressee is liable for the consequences, they have an incentive to take care of legally compliant AI decisions and to provide appropriate mechanisms. As a result, self-regulation could be achieved. In addition, it would also be possible to define a claim to human decision-making in appropriate cases, as in Art. 22 GDPR.

PRACTICAL CONSEQUENCES

From a practical point of view, it can be concluded from the ongoing discussion about the electronic person that liability issues are currently and will be subject to constant legal debate in the near future. All over the world, legislators are responsible for establishing regulations on who should be liable for autonomous decisions. This creates a higher degree of legal uncertainty for users, programmers, and adopters of AI.

These stakeholders need to pay attention to the directions in which regulatory trends will go. Although the legal concept of the electronic person appears to be a theoretically exciting approach, it is unlikely to find expression in legal regulations in the near future. This means that AI programmers will run the risk of being held liable for autonomous decisions in case of doubt.

SUMMARY

In conclusion, using AI remains operating in a legally gray area. Neither internationally nor in any jurisdiction is there a level playing field concerning AI. Foreseeable regulatory approaches, namely in the EU, remain far behind the need for regulation in the light of technological possibilities. Lawmakers should be more visionary when creating a legal framework for AI. The idea of an electronic person for sure is such a creative and visionary approach. However, it collides with basic legal principles. But in any case, the thought process of the electronic person in quintessence is correct. A legal framework for AI must strive, if it is to be visionary, for a clear starting point for automated decisions to be made. The person who acts must be as close as possible to the decision made by the AI.

In light of existing regulation and foreseeable legislative intent, users of AI should place a focus on the following legal requirements:

• Explainability and transparency of the functioning of the algorithms used.
• Observance of data protection requirements.
• Examine whether automated decisions can lead to discrimination.
• Comprehensive liability also for autonomous decisions.

REFERENCES

Armour, J., & Eidenmüller, H. (2020). Self-driving corporations? *Harvard Business Law Review, 10*, 87–110.

EY. (2019, May 17). *Künstliche Intelligenz spielt für Unternehmen große Rolle [Artificial intelligence plays a big role for companies].* https://cf-fachportal.de/meldungen/kuenstliche-intelligenz-spielt-fuer-unternehmen-grosse-rolle/

Isensee, J., & Kirchhof, P. (Eds.). (2003). *Handbuch des Staatsrechts der Bundesrepublik Deutschland* (3rd ed.). C.F. Müller Juristischer Verlag.

Kaulartz, J., & Heckmann, M. (2016, November 11). Selbsterfüllende Verträge: Smart Contracts: Quellcode als Vertragstext [Self-fulfilling contracts: Smart Contracts: Source code as contract text]. *c't, 2016*(24), 138.

Kingston, J. K. C. (2016). Artificial intelligence and legal liability. In M. Bramer & M. Petridis (Eds.), *Research and Development in Intelligent Systems XXXIII* (269–279). Springer International Publishing. https://doi.org/10.1007/978-3-319-47175-4_20

Lauwaert, L. (2021). Artificial intelligence and responsibility. *AI & Society, 36*(3), 1001–1009. https://doi.org/10.1007/s00146-020-01119-3

Manyika, J. (2017, October 2). *10 imperatives for Europe in the age of AI and automation | McKinsey.* www.mckinsey.com/featured-insights/europe/ten-imperatives-for-europe-in-the-age-of-ai-and-automation

Matthias, A. (2004). The responsibility gap: Ascribing responsibility for the actions of learning automata. *Ethics and Information Technology, 6*(3), 175–183. https://doi.org/10.1007/s10676-004-3422-1

Misselhorn, C. (2018). *Grundfragen der Maschinenethik [Basic Questions in Machine Ethics]* (4th ed.). Reclam.

Nathmann, M. (2021). Künstliche intelligenz im verfassungsrecht [Artificial intelligence in constitutional law]. In E. Flohr, M. F. Schmitt, & L. Gramlich (Eds.), *Vielfalt des Rechts: Festschrift für Ludwig Gramlich zum 70. Geburtstag.* C.H. Beck.

Ramge, T., & Galieva, D. (2022). *Mensch und Maschine: Wie künstliche Intelligenz und Roboter unser Leben verändern* (8. Auflage). Reclam.

Steinberg, R. (1998). *Der ökologische Verfassungsstaat [The Ecological Constitutional State].* Suhrkamp.

11

Industry 4.0 and Lean Production

Marc Helmold
IU International University of Applied Sciences, Berlin, Germany

Tracy Dathe
Macromedia University of Applied Sciences, Berlin, Germany

CONTENTS

DOI: 10.1201/9781003371397-11

PRINCIPLES OF INDUSTRY 4.0

Production systems in the twenty-first century are not like they used to be in the old days. Manufacturing companies are confronted with new generations of completely novel computer-based technologies in providing customer-centered products and services in the world market, with enormous competitive pressure (Bhamu & Singh Sangwan, 2014). To ensure long-term success, the companies need to adapt to shorter delivery time, increasing product variability and high market volatility, and react to continuous and unexpected changes sensitively and timely at reasonable costs (Moeuf, et al., 2020).

Industry 4.0 is commonly referred to as the 4th Industrial Revolution. This widely used term indicates the enormous impact of the current trend of automation and massive embedded data exchange with internet technologies on the contemporary industrial practice (see Figure 11.1).

Industry 4.0 in manufacturing is characterized by the processes of automized data collection and analysis, along with the information exchange through internet communication between the suppliers and the customers with regard to the products, the services, and the supply chain process (Buer

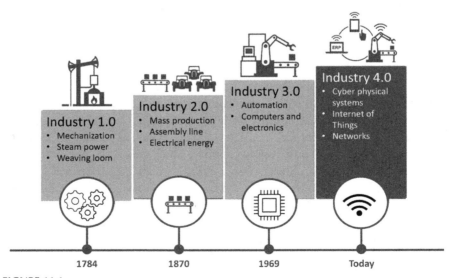

FIGURE 11.1
Industry 4.0 evolution. (Adapted from Industry 4.0: The Top 9 Trends For 2018.) (Kachur, 2018).

et al., 2018). In other terms, the digital information and communication technologies are implemented into production systems, which, as a result, combine the physical world with the advantageous fast data access and data processing technologies via the internet, including cyber-physical systems, the IoT, cloud computing, and cognitive computing.

Intelligent manufacturing in Industry 4.0 is often discussed with the key concept of "smart factory." A smart factory is established with a modular structure. CPS monitor the physical manufacturing processes, creating a virtual copy of the physical world to support decentralized decision-making. Over the IoT, CPS communicate and cooperate with each other and human operators in real-time, both internally and across organizations, as external

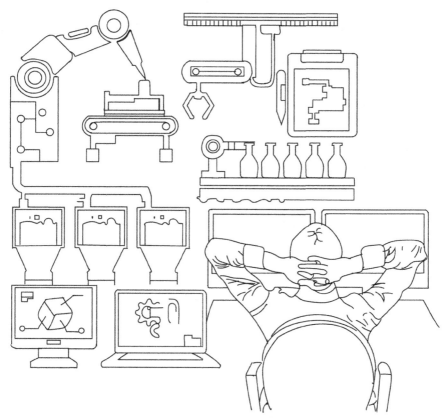

FIGURE 11.2
The production process is fully automated, and humans can relax.

value chain participants. There are four design principles in Industry 4.0 (Helmold & Terry, 2021):

- **Interconnection:** The ability of machines, devices, sensors, and people to connect and communicate with each other via the IoT or the IoP. The IoT paradigm focuses on the connections of nodes in production systems (objects and things) that combine the cyber and physical spaces, while the IoP paradigm aims at the human interactions in the cyber world (interconnection of human nodes) (Shi et al., 2021).
- **Information transparency:** The transparency created by Industry 4.0 technologies provides human operators with vast amounts of helpful information to make appropriate decisions. Interconnectivity allows operators to collect immense amount of data and information from all points in the manufacturing process, thus aiding functionality and identifying key areas that in turn benefit from more focused innovations and improvements.
- **Technical assistance:** First, the ability of assistance systems is to support humans by aggregating and visualizing information comprehensively for informed decision-making and solving urgent problems on short notice. Second, the ability of CPS to physically support humans by conducting a range of tasks that are unpleasant, too exhausting, or unsafe for their human coworkers.
- **Decentralized decisions:** The ability of CPS to make decisions independently and to carry out their tasks with a high degree of automation. Only in the event of disruptions or conflicting goals will the tasks be delegated to a higher level in the systems.

PRINCIPLES OF LEAN PRODUCTION

The term lean production is a management philosophy derived from the Japanese automotive industry. The main aim of LP is the achievement of customer satisfaction at a low cost of time and money (Kovács, 2020). In academic research, the LP practice is discussed in many various aspects of production systems, including just-in-time (JIT) flows, HR management, total productive maintenance and equipment management, and total quality management (Van Assen & De Mast, 2019). The JIT flow of LP systems can

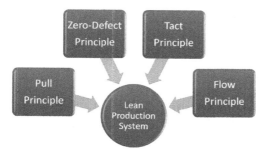

FIGURE 11.3
The four lean production principles.

be described as the ideal combination of four principles (Helmold & Samara, 2019; Imai, 1986) (see Figure 11.3):

- Zero-Defect Principle
- Pull Principle
- Flow Principle
- Tact Principle

Zero-Defect Principle

As the starting point in Toyota's success story, zero defects is all about identifying errors or defects as closely as possible to where they occur. By doing so, and by neither accepting nor passing on defects, problems are resolved quickly and efficiently, avoiding subsequent rework and quality issues.

The zero-defect principle is a concept of the Toyota Production System (TPS) to reduce defects through error prevention (Ohno, 1990). The basic idea is to motivate people to avoid mistakes by developing a constant, conscious desire to do their job right from the very beginning. Although the state of zero defects hardly exists, this concept ensures that no waste is caused in the production process (Helmold & Terry, 2016). Waste refers to all inefficient processes, use of tools, consumption of materials and staffing, as well as other resources. The process of elimination of waste is developed to avoid anything that is unproductive and does not add value to a project. Eliminating waste creates a cycle of improvement that increases cost efficiency. In industrial practice, the zero-defects theory is often interpreted as the concept of "doing it right the first time" to avoid costly and time-consuming corrections later in the project management process.

Pull Principle

This principle aims to avoid over-production and to avoid keeping unnecessary inventory, which in return saves working capital and reduces cash requirements.

In the traditional push system—the opposite of the pull principle—as many products as possible are manufactured and subsequently brought to the market through the sales channels. The mismatch between production planning and market demand often leads to suboptimal utilization of production capacities and storage space.

The pull principle, on the contrary, initiates the production process upon customer demand. Customer demand dictates the rate at which goods or services are delivered. Within the supply chain, components used in the manufacturing process are only replaced once they have been consumed by the upstream divisions. Thus, the pull system contributes to the waste reduction in the production process.

Flow Principle

Value creation should occur in a smooth, uninterrupted flow, from the start to the end of the production process. The ultimate effect of this principle is that all process steps are focused and geared toward adding value, one piece at a time, removing all wasteful and unnecessary activities from the process.

The benefit of a continuous flow in operations is that it features stability, continuity, balance, and no waste of time (the non-renewable resource). No time wasted on waiting between steps means time utility is maximized. Operations are not able to introduce a waste-less process without the continuous flow, as it is the true ideal process state. The ideal flow is the one-piece flow—some examples of such are shown in Figure 11.4.

Tact Principle

The term "tact" refers to the rhythm at which goods or services are produced to meet customer demand. Tact time is defined as the average production time

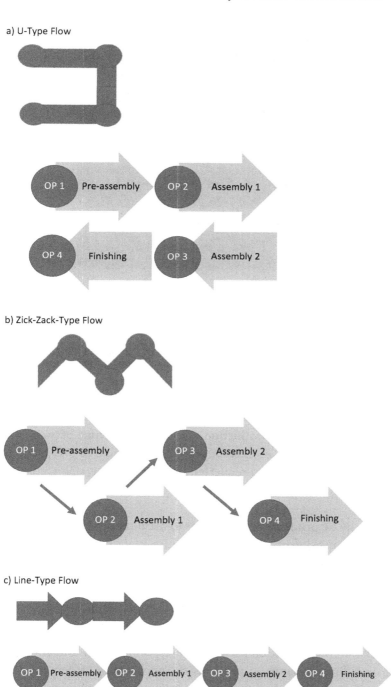

FIGURE 11.4
Types of flows in operations.

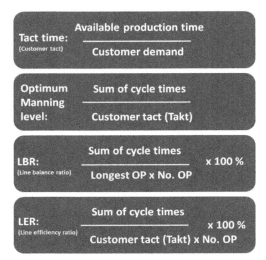

FIGURE 11.5
Tact time and other ratios.

available (total time available minus breaks, maintenance, or setups) divided by the production quantity requested by the customers (see Figure 11.5).

The average time between the start of production of one unit and that of the next unit should be set to match the rate of customer demand. For example, if a customer wants 15 units with the available time of 9 minutes and there is a steady flow through the production line, the average time between production starts should be 36 seconds for one unit (9 minutes = 540 seconds; 540 seconds divided by 15 unites requested by the customer = 36 seconds/unit). The ideal tact time has many benefits, for example:

- Directly tie production efficiencies to fiscal reporting
- Reduce investigation time for root cause analysis
- Shorten equipment return on investment through increased utilization
- Decrease costs through waste elimination
- Increase customer satisfaction through quality improvement

A consistent, continuous rhythm of production processes essentially facilitates the regulation of production activities to respond flexibly and effortlessly to rising or falling demand. In practice, the approaches of demand leveling, additional resources, or process re-engineering are often chosen to tackle tact time issues (Helmold & Terry, 2016).

A Selection of Toolsets for Lean Production

Andon

Andon (Japanese: アンドン or あんどん or 行灯) is a lean manufacturing tool referring to a system to notify management, maintenance, and other workers of a quality or process problem. The word Andon is a loan-word in English and means "lantern" in the original Japanese language. The center-piece of Andon is a device incorporating signal lights to indicate which work-station has a problem. The alert can either be activated manually by a worker using a pull cord or a button, or automatically by the production equipment itself. The system may include a means to stop production till the issue is solved.

An Andon system is one of the principal elements of the Jidoka method pioneered by Toyota as part of the TPS and, therefore, now part of the lean concept. It enables the worker to stop production when a defect occurs and they can then immediately call for assistance. Common reasons for manual activation of the Andon are part shortage, defect created or found, tool malfunction, or the existence of a safety problem. The alerts may be logged into a database so that they can be studied as part of a continuous improvement program. The system typically indicates where the alert takes place and may also provide a description of the trouble. Modern Andon systems can include text, graphics, or audio elements. Audio alerts may produce coded tones, for example, music with different tunes representing different types of disruptions or pre-recorded verbal messages (Imai, 1986).

Poka-Yoke

Poka-yoke (ポカヨケ) is a Japanese term that means "mistake-proofing." A poka-yoke is any mechanism in a lean concept that helps an equipment operator to avoid (*yokeru*) mistakes (*poka*). The concept was formalized, and the term was adopted by Shigeo Shingo as part of the TPS. It was originally described as baka-yoke, but as this means "fool-proofing" (or "idiot-proofing"), the name was changed to the milder poka-yoke.

The purpose of poka-yoke is to eliminate product defects by preventing, correcting, or drawing attention to human or other error sources. For example, a plug is designed with two different sizes of pins or three pins so that it can only be inserted into the socket in the right position.

Gemba and Shopfloor

Gemba (現場) is the Japanese term for the real or right place. A problem can be best solved at the location where it occurs. The management should frequently walk through the shopfloor—the most crucial place in the production environment—to discover the potential for improvement.

Shadow Boards

Shadow boards are specific boards for parts, tools, and equipment in operations, manufacturing, or service areas to reduce waste and waiting time. The shadow board helps to establish an organized workplace where inexpensive tools, supplies, and equipment are stored in appropriate locations where they are needed. It provides the basis for the standardization of the workplace.

The implementation of shadow boards helps to avoid waste, such as time looking for the appropriate tool or even having to buy a new one, wasted time in looking for supplies, and interchanging tools between tasks. The key is that they are appropriately located and hold all the necessary tools for the area or workstation. Shadow boards are also a subject of the 5S methodology and kaizen initiatives.

Kanban and Supermarkets

Kanban (看板) is a visual system for managing work as it moves through a process. Initially it arose as a scheduling system for lean manufacturing, originating from the TPS. Through the Kanban system, the production divisions are informed of what kind of products should be produced, as well as when and how much of those are needed.

Supermarkets are usually located near the supplying process to inform of customer usage and requirements. Each item in a supermarket has a specific location from which a material handler withdraws products in the precise amounts needed by a downstream process. Each time an item is removed by the material handler, a signal (such as a Kanban card or an empty bin) is sent to initiate the supplying process. Toyota installed its first supermarket in 1953 in the machine shop of its main plant in Toyota City (Ohno 1990). Toyota executive Taiichi Ohno took the idea for the supermarket from photos of American supermarkets showing goods arrayed on shelves by specific location for withdrawal by customers. Kanban and supermarkets are effective tools for the pull principle.

COMBINING INDUSTRY 4.0 WITH LEAN PRODUCTION

Lean AI Tools for Competitive Advantage

The LP approach is facing challenges of the growing complexity of the global market, especially in terms of requirements for rapid delivery, high degree of customization, and quick reaction to changing scheduling and product specifications (Buer et al., 2018; Moeuf, et al., 2020). The traditional decision-making based on human experience is driven to its limits in dealing with the mass information involved in production scheduling after the pull principle (Sanders et al.,2016). Increasingly, manufacturing companies are integrating Industry 4.0 technologies into their LP setups (Ciano et al., 2021). In particular, AI tools can lead to a competitive advantage across the value chain (Helmold & Samara, 2019) (see Figure 11.5). AI is a label for computer-aided systems capable of observing the environment, learning, and taking intelligent actions based on the obtained knowledge and experience to achieve predefined goals (Commission, 2018). Figure 11.6 depicts ten elements of AI tools that find wide applications in the industrial practice of LP (Helmold, 2021).

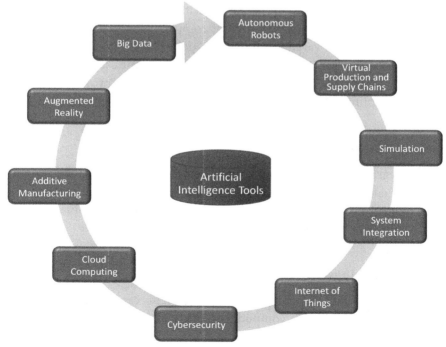

FIGURE 11.6
Artificial intelligence tools.

Autonomous Robots

An autonomous robot performs tasks with a high degree of autonomy (without external influence). Some common forms of autonomous robots are collaborative robots that perform tasks in collaboration with a human operator, automated guided vehicles that move items within sections of the supply chain such as production shop floors, warehouses, or distribution centers, as well as smart machines that communicate directly with other elements of the production systems (Ciano et al., 2021).

Virtual Production and Supply Chains

Virtual production elements are designed to visualize complex scenes or scenes that, due to the nature of the production, cannot be filmed for real. Virtual production can be implemented with filmic elements, typically with the aid of digital tools.

Lean Simulations

Simulation is used to project future development by mirroring the physical world in a virtual model and feeding it with real-time data. Lean simulation improves human understanding of complex production processes, the implications of input variables, and alternations of the value chain elements. Lean simulations can focus on design, manufacturing, capacity planning, or supply chain design. The simulation techniques can provide additional insights to optimize the production process.

Systems Integration

Lean integration is a continuous improvement methodology for bringing disparate data and software systems together to maximize customer value. Lean integration is a management system that emphasizes eliminating waste as sustainable data integration and systems integration practice.

Internet of Things

IoT in the context of production management can be understood as systems of heterogeneous computing devices and mechanical and digital machines that are interconnected with internet technologies to achieve predefined objectives. With their unique identifiers, the objects are enabled to communicate with each other by transferring data over a network without requiring human-to-human or human-to-computer interactions.

Cybersecurity

Cybersecurity means the protection of internet-connected systems, including hardware, software, and data, from cyberattacks. The modern security concept should comprise both cybersecurity and physical security in order to protect enterprises from unauthorized access to data centers and other computerized systems.

Cloud Computing

Cloud computing is a type of computing that relies on shared computing resources rather than having local servers or personal devices to handle applications. In its most simple description, cloud computing takes services ("cloud services") and moves them outside an organization's IT system and environment.

Additive Manufacturing

Additive manufacturing is the industrial production name for 3D printing, a computer-controlled process that creates three-dimensional objects by depositing materials, usually in layers. The official industry-standard term is the ASTM F2792 for all applications of 3D technology. It is defined as the process of joining materials to make objects from 3D model data, usually layer upon layer, as opposed to subtractive manufacturing methodologies.

Augmented Reality

AR is an interactive experience of a real-world environment where the objects that reside in the real world are enhanced by computer-generated perceptual information, sometimes across multiple sensory modalities, including visual, auditory, haptic, somatosensory, and olfactory.

Big Data

Big Data is a phrase used to mean a massive volume of both structured and unstructured data that is so large it is difficult to process using traditional database and software techniques. In most enterprise scenarios, the volume of data is too big, it moves too fast, or it exceeds the current processing capacity. The IoT technologies in various industry sectors inevitably lead to a dynamic increase in data volume. Thus, big data technologies have become indispensable for efficient data processing.

A CASE STUDY – SMART MANUFACTURING WITH INTERNET OF THINGS AT INTEL

Intel Corporation is an American-based multinational technology company. The largest semiconductor chip manufacturer operates in 46 countries with around $80 billion in annual sales revenues (Intel, 2022). In the recent years, Intel has reported sustainable success in manufacturing management by combining big data and IoT technologies, predominantly in the aspects of:

- Cost reduction
- Production efficiency
- Production quality

An internal white paper of the Intel factory management demonstrates the combination of Industry 4.0 and LP. The management primarily aims at minimizing manual processes with the target of complete automation. Over time, the implementation of cutting-edge technologies continuously intensifies the automation of the production processes, supported by a computer-aided, data-based decision-making system, which in turn significantly improves the competitive advantages and value creation of the business.

The optimization concept is built on a robust IT infrastructure that mainly includes (see Figure 11.7 Intel: the IT-infrastructure):

- The data storage and processing network, including servers, data storage devices, and standard software and hardware protocols.
- Sensors on the factory floor as terminals of the data processing network (IoT). Using sensor terminals like radio frequency identification tags, the real-time flow of materials (e.g., raw material, finished products, and work-in-process) can be monitored as they move along the production process.

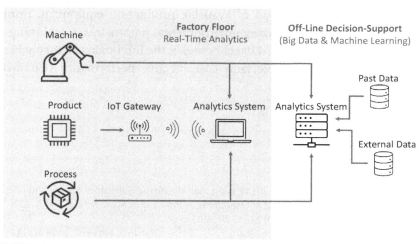

FIGURE 11.7
Intel: the IT-infrastructure.

- Integrated data analytics (system integration). The manufacturing execution system was created to register the current state of each production step at each equipment permanently so that the production transactions can be optimized, largely automatically by the predefined algorithms based on the shared database.
- Decision support tools based on big-data technology. The large quantity of collected data on the products, the machinery, and the production processes are further processed in an offline analytics system. The current data are combined with historical data to project future trends.

The real-time process control of the smart factory relies on the communication between the production equipment and the analytics system on a 24/7 basis. The material flow and the machinery performance are tracked with the sensors and the standard interfaces. Using the real-time data inputs and the predefined statistic thresholds, the factory floor analytics system's decision-support component automatically determines incremental process adjustments. In case of statistically significant deviations, material flows will be rerouted to redundant production equipment, and human and technical aids will be notified to examine the devices in question.

In addition, the offline analytics system uses historical and additional external data to project future trends and identify potentials for improvements, especially in terms of accelerating throughput, reducing time-to-market, increasing effective workload of available production equipment, minimizing error rate, and optimizing environmental parameters. Combining the LP concept and the Industry 4.0 technologies, the Intel smart factory achieves sustainable, resilient, flexible, and cost-effective performance (Chadwick, Lee, Meyer, & Sartini, 2016).

REFERENCES

Bhamu, J., & Singh Sangwan, K. (2014). Lean manufacturing: Literature review and research issues. *International Journal of Operations & Production Management*, 34(7), 876–940. https://doi.org/10.1108/IJOPM-08-2012-0315

Buer, S. V., Strandhagen, J. O., & Chan, F. T. (2018). The link between Industry 4.0 and lean manufacturing: Mapping current research and establishing a research agenda. *International Journal of Production Research*, 56(8), 2924–2940.

Chadwick, S., Lee, D., Meyer, S., & Sartini, J. (2016, April). https://media16.connectedsocialme dia.com/intel/04/14304/Using_Big_Data_Manufacturing_Intel_Smart_Factories.pdf. Retrieved from Intel (White Paper): https://media16.connectedsocialmedia.com/intel/04/14304/Using_Big_Data_Manufacturing_Intel_Smart_Factories.pdf

Ciano, M., Dallasega, P., Orzes, G., & Rossi, T. (2021). One-to-one relationships between Industry 4.0 technologies and lean production techniques: A multiple case study. *International Journal of Production Research*, 59(5).

Commission, E. (2018). *JRC Publications Repository* . Retrieved from https://publications.jrc.ec.europa.eu/repository/handle/JRC113826

Helmold, M. (2021). *Kaizen, Lean Management und Digitalisierung. Mit den japanischen Konzepten Wettbewerbsvorteile für das Unternehmen erzielen.* Wiesbaden: Springer.

Helmold, M., & Samara, W. (2019). *Progress in Performance Management. Industry Insights and Case Studies on Principles, Application Tools, and Practice.* Heidelberg: Springer.

Helmold, M., & Terry, B. (2016). *Global Sourcing and Supply Management Excellence in China. Procurement Guide for Supply Experts.* Singapore: Springer.

Helmold, M., & Terry, B. (2021). *Operations and Supply Management 4.0. Industry Insights, Case Studies and Best Practices.* Cham: Springer.

Imai, M. (1986). *Kaizen. Der Schlüssel zum Erfolg der Japaner im Wettbewerb.* Frankfurt: Ullstein.

Intel. (2022). *Intel Reports Fourth-Quarter and Full-Year 2021 Financial Results.* Retrieved from press releases: www.intc.com/news-events/press-releases/detail/1522/intel-repo rts-fourth-quarter-and-full-year-2021-financial

Kachur, L. (2018). *Industry 4.0: The Top 9 Trends for 2018.* Retrieved from AI Zone: https://dzone.com/articles/industry-40-the-top-9-trends-for-2018

Kovács, G. (2020). Combination of lean value-oriented conception and facility layout design for even more significant efficiency improvement and cost reduction. *International Journal of Production Research*, 58(10), 2916–2936.

Moeuf, A., Lamouri, S., Pellerin, R., Tamayo-Giraldo, S., Tobon-Valencia, E., & Eburdy, R. (2020). Identification of critical success factors, risks and opportunities of Industry 4.0 in SMEs. *International Journal of Production Research*, 58(5), 1384–1400.

Ohno, T. (1990). *Toyota Production System. Beyond large Scale Production.* New York: Productivity Press.

Sanders, A., Elangeswaran, C., & Wulfsberg, J. P. (2016). Industry 4.0 implies lean manufacturing: Research activities in Industry 4.0 function as enablers for lean manufacturing. *Journal of Industrial Engineering and Management, 9*(3), 811–833.

Shi, F., Wang, W., Wang, H., Ning, H., & Ning, H. (2021, April). *Research Gate.* Retrieved from The Internet of People: A Survey and Tutorial: www.researchgate.net/publication/350808415_The_Internet_of_People_A_Survey_and_Tutorial

Van Assen, M., & De Mast., J. (2019). Visual performance management as a fitness factor for lean. *International Journal of Production Research, 57*(1), 285–297.

12

Open Banking and Digital Ecosystems

Martin Schneider

Landesbank Hessen-Thüringen, Offenbach, Germany

CONTENTS

THE FUTURE OF DIGITAL BANKING

Digitization has a major impact on banks and customers. At the same time, the financial industry must respond to changing customer needs and new customer behavior by serving customers with innovations and changed products and services. Some banks find it difficult to modernize and digitize the IT infrastructure and system landscape within the banks. Nevertheless, the trend will intensify, which will lead to the implementation of new business models and increased cooperation and collaboration.

DOI: 10.1201/9781003371397-12

WHAT IS OPEN BANKING?

Key takeaways of The Investopedia Team (2020) are that:

- "Open banking is the system of allowing access and control of consumer banking and financial accounts through third-party applications.
- Open banking has the potential to reshape the competitive landscape and consumer experience of the banking industry."

Open banking is the opening of services and infrastructure of banks to non-banks. Some banks define it as providing application programming interfaces

FIGURE 12.1
Connecting systems as well as connecting providers and consumers with services and products.

(APIs) that are publicly available to authorized non-banks so that these can access the bank data, processes, and business functions via those interfaces. Sometimes it is also defined as the controlled opening of standardized interfaces for the secure exchange of data between banks, insurance companies and FinTechs, private clients, corporate clients, and non-banking partners. (Saritha, 2021)

At the same time, open banking can be much more: A platform-based business model in an ecosystem of banks, customers, third-party developers, FinTechs, and other partners in which new services can be created from banks as well as others. The financial data is accessible to various banks and FinTechs via these open banking APIs. Every party in the ecosystem can develop innovative services for customers so that bank customers get customer-centric services and applications from banks and non-banks based on open banking APIs and adapted for customers' needs.

Ecosystems develop from platforms and are an evolutionary step further, as they are supposed to be a framework for business relationships of shared value creation with higher-level objectives. A holistic customer-centric approach, in which the banking services are networked with those of partners and complemented by innovative services outside of banking, gives the customer a new customer experience.

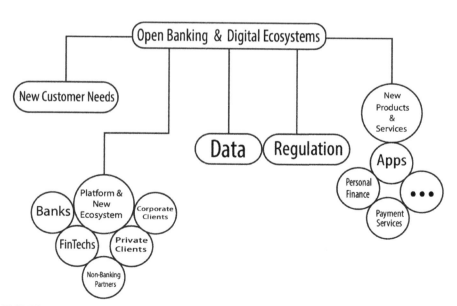

FIGURE 12.2
Open banking and digital ecosystems.

It is important to mention that the data is only exchanged if there is consent from the customer. That this, exchange is done in a standardized and secured manner between the bank and a trustworthy third-party provider. The participants of an ecosystem decide for themselves how they integrate third parties and what access options are granted to the partners to bank systems and what data is shared if there is consent by the customer. Different standards are applied to this situation, and it seems possible to use not only industry standards of API but also non-mandatory standards, which are not required to be a subject of regulation.

WHAT DOES OPEN BANKING MEAN FOR THE CUSTOMER?

Today's customers have changing needs, and they can choose between many providers in the digital market, which results in the adaptation of corresponding offers to the customer's needs. For customers, open banking means a better payment experience, such as extended service (e.g., for tax returns or through purchasing discounts), as well as automation through integration into the IT landscape of other providers such as online shops or account aggregation; however, the consent of the customer is always a prerequisite for open banking activities. Two brief examples of how open banking can help private and corporate customers are presented in the following text. A private customer can access all the bank services; for example, account details in one place, even if the customer has accounts at several different banks. The customer can also use an application by a non-bank to access the accounts.

Another use for a corporate customer can be the connection between the customer's accounting program directly to the various bank accounts so that incoming and outgoing payments are controlled and mapped on the bank accounts in the accounting program. The customer can oversee everything in one dashboard.

CHRONOLOGY OF OPEN BANKING

The importance of traditional banking was very high in the past because it would take place offline and customers had to come to the bank branch to find out the current account balance, make transfers, or speak to the client advisor, who could offer the products and services by the respective bank.

Online banking has been around for several years already, and mobile banking is on the rise or is already prevalent among many private customers. With these new channels, consumers now have the opportunity to do many banking transactions digitally. Physical bank branches still exist in some cases, mainly used for more complicated banking operations or by customers who reject the digital channels.

In online or mobile banking, the bank's information about the customer or banking products is provided via a frontend user interface. The user interface system accesses the banking systems via interfaces; for example, a mobile banking app is getting data on the account balance of the customer to show in the banking app. Some of these interfaces are available to authorized providers so that they can offer their own products and services in addition to the digital offerings of the banks inside the app, which led to more transparency and opening up of the banks during the last couple of years.

Third-party service providers such as regulated FinTechs are provided with customer data and transaction data via these open banking interfaces to make payment transactions if bank customers agree to it. Consumers can make payments, for example, via a mobile app from a third-party service provider frontend. Customers who have accounts with multiple banks can also bundle in such a third-party service provider app or view the accounts of various financial institutions in online banking or mobile banking at a bank and make payments.

This allows customers to take advantage of the offer of many other providers. Not only are they limited to one bank, but they get services and products from several providers and can look for the best service for themselves. In the future, this trend will continue and customers will take advantage of offers across company and without boundaries.

In the EU, the UK, and many other countries, through legislation, directives, or guidelines, the opening up has accelerated considerably. However, the opening of banking (i.e., open banking) is not distributed worldwide on the same level. For example, in the EU and Brazil financial institutions are required to open account information and payment trigger services to third parties. In Australia it is the same and, in addition, the banks there must also disclose product information and service fees. On the other hand, the third-party providers who can access this information are also regulated and customer data is not available to everyone openly. Then there are countries, such as Hong Kong, Singapore, or Japan, where regulators encourage the exchange of data between financial institutions and third-party service providers but do not impose this on financial institutions. In addition, there is a market-based approach in several countries, where the regulator does not make any specifications, but market players have recognized the benefits of open banking and, without explicit rules for data exchange, operate it in the appropriate style for them. This can be seen in Switzerland, China, and the USA. (Aytas et al., 2021; Bank for International Settlements, 2019).

Banking as a service is on the rise. Banking as a service means the bank does not have the primary customer relationship. Still, the customer first becomes aware of the bank's offers via one of the third-party service providers because it is the right offer or service for the customer at this very moment. However, banking as a platform is another evolutionary step, while new ecosystems are created, and the consumer can access a variety of offers on platforms, partly from the financial sector but also from other areas, and the consumer decides which provider she/he wants to contact and work with. The customer thus uses the offers across company boundaries. These platforms are not subject to financial supervision, so they are not regulated.

FINTECHS AND THEIR BUSINESS MODELS

A clear definition of the term FinTech does not exist currently. The word FinTech is made up of the words financial services and technology and is understood to mean young technology-oriented companies with customer-oriented financial services. The business models of these FinTechs can be very heterogeneous. According to the German Federal Financial Supervisory Authority (n.d.) the business models of FinTechs are mainly, but not conclusively:

- alternative payment methods
- automated financial portfolio management
- blockchain technology
- crowdfunding, crowd investing, crowd lending
- platforms for automated investment advice
- platforms for signaling and automated job execution
- virtual currencies
- InsureTechs, FinTechs in the insurance industry

The mentioned new business models of FinTechs are sometimes realized without, but often with, cooperation with traditional partners from the financial industry. This is because the great strength of FinTechs often comes from working with other business partners such as larger banks. Both banks and FinTechs have their respective strengths. FinTechs, for example, are often much more flexible and can respond quickly to customer reactions. Banks, on the other hand, have very high security standards and often hold customer relationships.

DRIVERS OF CHANGE

There are various reasons that have accelerated this rapid development. Open banking serves the new customer needs. Among other things, customers want to be able to use tailor-made products and their banking business at any time and from anywhere. Customer behavior is changing constantly and faster, with customers recognizing that they have the power to decide which financial institution they want to work with. In addition, customers regain control of their own data by being allowed to decide whether and with whom they want to share the data. The financial industry is becoming increasingly technology-driven and new opportunities are emerging for both customers and the banks.

Competitive pressure has increased significantly. There are also new market participants, such as FinTechs, which have taken over parts of the market previously granted to the classic banks. However, new opportunities also arise for the traditional banks; for example, through cooperation with others or by exploiting new markets for the banks. In addition, large non-banks such as Google or Facebook are pushing into the market and putting pressure on traditional players. These large technology companies adapt their business models quickly and with customer focus and understanding of how to offer innovative products and services. In addition, many banks are struggling with the ongoing low-interest-rate environment and cost increases.

Additionally, suppliers and consumers are using new value chains. Banks and non-banks are networking, and ecosystems are emerging that are supported by new API technologies. There are now so-called API kits and API standards that allow participation in this market very quickly and easily. There are also other technology trends that are accelerating the opening of traditional banks and banking, such as distributed ledger technology or so-called blockchain or banking-as-a-service facilities via cloud technology.

The requirements of the regulatory authorities and the new standards that they introduce are also a curse for some market participants because they are struggling to implement them. For others, including new market participants, they are a blessing and enable them to be part of the market. Among other things, this is also about the standardization and cooperation of the systems. This is sometimes very problematic for a highly complex financial company with a grown structure, whereas lightweight startup companies in the financial sector have it much more manageable. Not only is the customer data protected better and uniformly by regulatory systems such as GDPR, but also through payment service directives a liberalization of the data has been initiated, which is progressing inexorably.

NEW BUSINESS MODELS IN THE FINANCIAL INDUSTRY

Open banking provides the opportunity for consumers, bank customers, banks, as well as FinTech to provide their products and services. The new holistic customer experience cannot be achieved by a single bank or financial institution but must be achieved through complementary partnerships. Open banking is an enabler technology like the internet was in the past. That is to say that there are not only emails on the internet but, through the internet, there are an infinite number of useful offers for a wide variety of customers.

Open banking offers the opportunity for new business models in the financial sector because customer data is made available and suitable offers for the consumer are created. At the moment there are already improved payment services. There is, for example, a request to pay at the end customer business, which will considerably replace the advance payment. The payee requests money from the customer by means of a submitted payment request. The customer does not have to enter or fill in anything. They only need to confirm the payment for the payment to be made. This is much more convenient than traditional bank transfer prepayment. In addition, consumers can take advantage of additional services from different third-party providers and still have their own chosen channel of entry via their own trusted bank.

The opening of the API platform to other players in the financial sector arises when, for example, insurance companies, FinTechs, or InsurTechs connect their platforms via open banking APIs. A digital ecosystem arises when the platforms of each sector are networked via open banking APIs. A platform of insurance services is linked to banking services and the retail platform as well as other platforms or players from other sectors, such as automotive companies. The collaboration between several players enables the different capabilities and thereby generates customer benefits. First of all, banks offer open banking APIs to help them sell their banking products and banking services to third parties. In addition, the banks also enable third-party providers to integrate their services into the bank's platform. Innovative services are offered to customers through access to these API platforms by banks focusing on customer needs and user experience in order to offer added value to the customer in addition to their own banking products and services. With the partnerships in platforms and with companies, the customer benefits increase in a targeted manner (Farrow, 2020). One wonders which parts of the value-added are to be taken over by the banks themselves and how one can maintain direct interaction with the customer.

Open banking APIs extend the offering of the classic bank and increase the customer experience. Through the platform or the ecosystem, partnerships are established with FinTechs and other third-party providers such as payment service providers. This provides banks with additional and new revenue opportunities and higher value-added. In addition, the innovative strength of the banks will be greatly increased. In this way, banks optimize their distribution model by combining with selected ecosystems and expanding their partner network by integrating open banking API. They also increase employee productivity by automating processes or streamlining workflows and eliminating tasks and issues.

Open banking will become increasingly important. Financial institutions and other market participants will try to meet the new market conditions and customer needs and will invest in digitizing their business models. New digital ecosystems will emerge, and the boundaries between the financial industry and other areas will vanish as third-party providers from different sectors will participate in the digital ecosystems to meet customer expectations.

DATA PROTECTION AND SECURITY

Open banking has many benefits, but the risks also need to be considered. Above all, there are concerns related to personal data protection and cyber security. For all participants in open banking, it is of fundamental importance that data transmission takes place securely and appropriately. Secure access, appropriate authentication, and decent use of the available data play an important role here. Therefore, data storage and transmission must be encrypted, and data ownership must be defined. It is also important that data privacy standards are observed, and governance and risk management takes place regularly and properly. When opening APIs, security and penetration testing must be carried out in order to find and eliminate vulnerabilities. It is imperative that banks retain the trust that they have from their customers, and it is crucial that customers understand how their data is used and that they are given the opportunity to grant or deny access to their data. (Murphy, 2017; Nanaeva et al., 2021; Schaus, 2019).

SUMMARY

Open banking products and services are characterized by the fact that they are created in cooperation with a third party and complement the traditional banking offering. By transforming their business model and adopting open banking, banks are encouraging collaboration to find new ways to respond to new customer needs and serve customers in the best way possible. In addition, banks face the challenge of new technologies, highly competitive pressure and novel regulatory requirements and standards. Through the efficient use of open banking and the creation or participation in ecosystems, banks differentiate themselves from their competitors through the connected services of third-party providers for their customer target group.

REFERENCES

Aytas, B., Öztaner, S. M. & Sener, E. (2021). Open banking: Opening up the "walled gardens". *Journal of Payments Strategy & Systems, 15*(4), 419–431.

Bank for International Settlements. (2019, November). *Report on open banking and application programming interfaces.* Retrieved January 20, 2022, from www.bis.org/bcbs/publ/d486.pdf

Bundesanstalt für Finanzdienstleistungsaufsicht. (n. d.). *Unternehmensgründer und Fintechs.* Retrieved March 2, 2022, from www.bafin.de/DE/Aufsicht/FinTech/finte ch_node.html

Farrow, G. (2020). Open banking: The rise of the cloud platform. *Journal of Payments Strategy & Systems, 14*(2), 128–146.

Murphy, K. P. (2017). Open banking is inevitable. Let's rethink data security, too. *American Banker, 182,* 99.

Nanaeva, Z., Aysan, A. F. & Shirazi, N. S. (2021). Open banking in Europe: The effect of the Revised Payment Services Directive on Solarisbank and Insha. *Journal of Payments Strategy & Systems, 15*(4), 432–444.

Saritha, M. (2021). Open banking in India: A technology revolution in the banking sector. *IUP Journal of Accounting Research & Audit Practices, 20*(4), 572–577.

Schaus, Paul (2019). When open banking and data privacy collide. *American Banker, 184,* 238.

The Investopedia Team. (2020, August 27). *Open Banking.* Retrieved November 21, 2021, from www.investopedia.com/terms/o/open-banking.asp

13

New Self-Managing Work Systems

Kourosh Dadgar
University of San Francisco, San Francisco, California, USA

CONTENTS

NEW WORK MODELS AND TECHNOLOGY – ENABLED WORK SYSTEMS

On April 28, 2022, Airbnb announced that it will allow employees to live and work anywhere (Brian, 2022). The company plans to attract and promote even more remote works. Their "live and work anywhere" policy has five key features: You can work from home or office; you can move anywhere in the country you work in and your compensation will not change; you can travel around the world and can live in over 170 countries for up to 90 days; today more than 20 countries offer remote work visas; there will be regular in-person meet-ups, team gatherings, and social events. This flexible work system is executed based on a single company calendar with product releases in May and November to maintain company-wide alignment. The CEO of Airbnb in his message to the Airbnb employees reiterates that this is the future of work ten years from now, a new work model that the world will embrace in 2022 expedited by the 2019 pandemic.

DOI: 10.1201/9781003371397-13

McKinsey predicts that post COVID-19 pandemic, 25% of the workers need to switch jobs and remote work will be adopted more by companies (Lund et al., 2021). McKinsey's investigation of labor, workforce, and jobs for eight industrial countries provides interesting insights about the characteristics of the Industry 4.0 new works. It is important to know what type of jobs will change and adapt to the new work models. McKinsey groups jobs into ten categories based on the physical proximity required in those jobs. Proximity at work is characterized by proximity to coworkers and customers, number of interpersonal interactions, and on-site and indoor activities. Jobs with higher levels of proximity are more likely to change and adapt to new work models enabled by automation and AI technologies and systems. For example, in medical care and personal care, on-site customer interaction with the highest levels of proximity are more likely to change into tele-health, remote personal tele-trainers, and computer-mediated communication with the customers. Remote work and virtual meetings will be more prevalent post-pandemic but jobs that require rich in-person human interactions and presence, such as negotiations and onboarding new employees, will lose their quality if they are done remotely.

The new work model in the U.S. was actively employed in education, healthcare, transportation, and retail industries during the pandemic. In education, HyFlex (Hybrid-Flexible) (HyFlex Learning Community, 2022), remote, and hybrid classes have become more prevalent in U.S. colleges and universities. The HyFlex classes help students join classes remotely without compromising their learning experience. In healthcare, physicians visit their patients electronically through tele-health services unless an in-person physical checkup or examination is necessary. In transportation, a few companies are making breakthrough progress to move riders to their destinations in autonomous vehicles. The autonomous vehicle maker in San Francisco offers driverless services to the residents of San Francisco (Cruise, 2022). And in retail industry, online and internet-based shopping experiences are exponentially growing and changing the consumer behavior and shopping preferences. These changes are being adopted in other industries as well.

New work models are more inclusive. Works are democratized in the new work models and people of different educational and ethnic backgrounds are less marginalized. New smart devices, sensors, and advanced analytical and data-sharing technologies create sharing economy capabilities that allow business to distribute works and optimize workflows. Ridesharing companies like Uber and Lyft in the U.S. are the best first examples of such businesses in the world. If we compare Uber, the ridesharing business headquartered in the U.S., with the traditional taxi services, we can highlight a few pros and cons. Rideshare works are available to a larger number of people with diverse educational and socioeconomic status. Any individual who owns a car can be potentially employed by the rideshare companies if they pass routine hiring criteria. Individuals hired by these companies can work around flexible schedules beyond fixed working hours and gain an additional income. Part-time flexible works do not qualify for healthcare and benefits that are usually provided for full-time works, which poses challenges for new workers and companies. Work-life balance will be different in the new work models.

It is expected that new works open up some time for workers to attend to their lives but, in reality, low pay and high demands require workers to extend their working hours to fill in the pay gaps and satisfy the expectations of the companies. These issues are beyond the scope of this chapter but certainly should be addressed. New technology-enabled work systems have extended into adjacent similar industries, influencing their ecosystems. Rideshare work systems have penetrated and changed public transport and auto rental markets. Uber and other rideshare services have integrated taxi and other public transport modes into their platforms and they have expanded their services into the auto rental industry as well. This exponential growth enabled by advanced technologies will permeate into all industries and markets and will permanently change their work systems and flows. The changes in the work systems are focused on the delivery of a product or a service. In education it is teaching and learning, in healthcare it is delivery of care to patients, in transportation it is moving the customers from point A to point B, and in retail industry it is selling and delivering the product to the customers worldwide. The focus on the efficiency of product and service delivery requires changes in the works and processes that enable it.

Companies and employees learned during the pandemic that not all jobs and activities need to be done in person and organizational efficiency and individual productivity can significantly improve when done remotely. Companies and new workers are required to adapt to the new work systems and change their working habits and mindset so that they can function effectively and efficiently within these systems and avoid life disruptions, and mind and body exhaustion. In the next section, work system framework is introduced to conceptually frame and explain the complexity of work models and systems and their interplays.

WORK SYSTEM THEORY AND FRAMEWORK

In organizations, work is the use of resources, such as informational, human, physical, to produce and deliver a product or a service, and a work system is a system in which humans and/or machines do the work, consisting of process and activities using information, technology, and other resources, to produce specific products or services for customers (Alter, 2013).

Work systems are ICT-reliant but the role of ICT in them varies across different industries. There are different instantiations of work systems. A work system can be the whole organization or it could be a work practice. For instance, in healthcare, there is a diabetes management work system in which patients with diabetes use devices such as glucometers and insulin pens to manage their diabetes by using diabetes information such as glucose levels, calories and carbohydrate amounts in food, in conjunction with endocrinologists, nurses, nutritionists (Dadgar & Joshi, 2018).

Work system theory proposes a new framework that focuses on ICT-enabled systems in organizations in which customers are not merely technology users but active components of a system in the organization that produces desired business results in the form of products and services (Alter, 2013). Technology-enabled platforms and processes have turned customers and clients in different industries into active participants involved in the production of the final product or service. For instance, in healthcare, active care models are created in which patients actively participate in their care process in conjunction with their healthcare providers by collecting and using their disease and treatment data to manage their health conditions. Work systems help organizations maintain an effective strategic alignment between the works that are performed, the information and technologies that are used, and the final product and service that is delivered.

New work models are characterized by the lack of direct leadership and immediate resources that are evident in the remote works and flexible working hours. It is essential for the new workers to effectively self-manage their works and perform processes and activities. In the next section, self-management and self-managing workers are discussed, and their structure, characteristics, and contingencies are explained.

SELF-MANAGEMENT AND SELF-MANAGING WORKERS

Avogaro (2019) reports that smart and mobile technologies enable iterative interplays and interactions between research, production, services, and consumer works. In this new intelligent and iterative work system, businesses know about their customers' preferences and needs in real-time, which in turn leads to a JIT production based on customers' demands and moving market needs. Remote flexible works can facilitate real-time and efficient data-driven decision-making required for such business relationships between consumers and production of services and products. Against this backdrop of Industry 4.0 new work models, workers are expected to take different roles and effectively use their non-routine cognitive skills, such as critical thinking and problem-solving, rather than limiting themselves to strictly following their respective supervisors and managers. New workers, who need to be trained and educated so that they can adapt to such an iterative work dynamic, will need to be more creative and independent in the absence of direct instructions, supervision, and resources to self-manage their job responsibilities and activities.

Self-management is defined as behavior formed in the absence of immediate resources and constraints (Manz & Sims, 1980). Self-management is characterized by how employees in organizations respond to the lack of constraints and leadership, and what consequences and reinforcements they experience based on their responses. In the lack of constraints and resources, employees face different response alternatives and consequences. Responses are evaluated against a criteria and reinforcements are provided accordingly. The management team that evaluates responses and determines reinforcements is critical in maintaining an effective long-term self-management in the absence of leadership and immediate resources.

The structure through which an effective self-management is executed is important. Ineffective self-management may be attributed to internal individual attributes (such as lack of hard work or intellectual capacity) or external factors (such as the organizational structure or workflows). An effective organizational work structure limits inefficiency that may negatively influence individual efficiency and performance of the self-managing individual workers. An effective self-management offers benefits to the organization and its employees. Self-managed works open up some quality time for the organization and save time and cost in performing necessary organizational processes and activities, which allows the organization to focus on long-term pressing problems. A self-management governing body needs to ensure self-managing individuals work toward realistic goals and avoid dysfunctional self-management. Self-management governance of works facilitates an effective self-management. The self-management part of the organizational governance determines the evaluation criteria and reinforcements for the self-managed works.

The new work models rearrange different components of a typical work system by Alter (2013) and explain the central role of technology and the importance of and reliance on data and information to perform processes and activities. Such changes are inherent and inevitable in the new work models that workers self-manage their works. In the next section, work system framework is adapted to explain the new self-managing work systems and models.

NEW SELF-MANAGING WORK SYSTEMS

The works that can be disrupted and changed in part or completely into technology-enabled remote works or technology-assisted reduced works need to be self-managed by the workers. These new self-managed works need to be evaluated, reinforced, and structured to achieve effective self-management. Self-management in the absence of direct and immediate leadership and resources is managed based on the responses of the workers to problems and challenges and the reinforcements and consequences that will help leadership indirectly evaluate them. Access to informational resources plays a critical role in forming and defining how workers respond effectively and choose alternatives that are strategically aligned with the organizational objectives. Post-digital revolution information is generated in abundance and stored, shared, processed, and analyzed inclusively and optimally across all industries, works, platforms, and practices. New workers can effectively self-manage their work responsibilities based on necessary information if they are equipped with non-routine cognitive skills such as critical thinking, problem-solving, and analytical thinking.

The main reinforcement for the new works is inherent in its form and that is having flexible working hours and location. New workers in ideal cases should be able to better balance work and life and the performance expectations should be reasonably balanced. This inherent attribute of the new works reinforces new workers' behavior to effectively self-manage their activities and responsibilities within the framework of the remote work and keep the company's trust to perform the work in the absence of direct leadership and immediate resources. The structure through which new works are performed should enable effective self-management of works and responsibilities, be flexible with easily accessible informational resources and outcome-driven expectations, and provide a mechanism and platform to evaluate works. The new self-managing work systems provide such work structures.

New self-managing work systems are described and modeled against the work system framework (Alter, 2013) and they consist of the basic components of work systems rearranged based on the Industry 4.0 standards and capabilities (Figure 13.1). The new self-managing work systems are information-intensive iterative structures in which business and customer participants are data workers who make data-driven decisions and use data to perform works and meet their responsibilities. Customers use data to measure the quality and functionality of the products or services that are delivered to them. Sometimes customers are directly involved in the service delivery processes and activities, such as patients who actively work

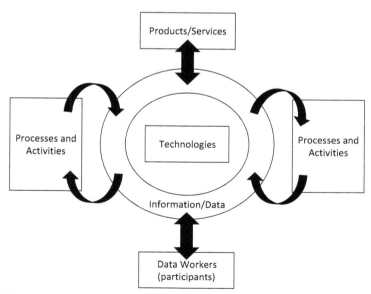

FIGURE 13.1
Self-managing work systems.

with their care providers to improve the quality and effectiveness of their treatments and healthcare condition. Access to information and digestible and intelligible data are the conveniently and easily accessible and available source of decision-making and work execution.

Technologies facilitate and enable data-driven execution of processes and activities necessary to deliver the final product or service. Post digital revolution, cheap and fast technological capabilities have become available for the organizations to collect, store, and analyze data required to perform organizational processes and activities such as product development and project management. Digital revolution has made organizations and their operational units and departments more connected and collaborative and it has fundamentally changed how works are done and has disrupted their structure and nature. Data workers constantly self-manage their works with immediate and convenient access to relevant and necessary information throughout the works.

Organizations must focus on the essential components of self-managing work systems to prepare for the new works or enhance their existing work structure and flows to enable effective self-management that will be done in the absence of immediate and constant leadership. Self-management may be done in conjunction with higher management if the works require direct and continuous collaboration between the workers and leadership. Technology plays a central role in the work systems to allow and enable data-driven works, processes, and decisions. Advanced technologies enable digital platforms to efficiently store, share, and process structured and unstructured data between the workers and management team. Efficient data-sharing platforms effectively enable self-managing, data-driven works. Data quality is an important factor in ensuring the basis of quality operational, managerial, and strategic decisions. Lower data quality will result in products and services that deliver lower qualities to the customers and clients.

New self-managing work systems are implemented based on a flexible organizational structure that facilitates and maintains an effective self-management and allows the management team to evaluate the works and reinforce the workers' performance in the absence of immediate resources and direct and constant leadership and supervision. Such self-managing structures should streamline data sharing across the work platforms, including communicative technologies such as Zoom (Zoom Video Communications, Inc., 2022) that is the prevalent platform in the U.S. to hold remote meetings, training, reporting, and presentation sessions. Continuous governance of such work systems is necessary to ensure that they are self-sustaining and in the case of crisis and emergency they are adaptive to changes such as the pandemic and technology infrastructure disruptions.

SUMMARY

Post digital revolution, new technological advancements in AI, machine learning, and computational capacities to share, store, process, and analyze large amounts of data led to the emergence of new types of work systems and structures. New work systems enabled by advanced technologies are data-driven and self-managing. In the absence of immediate resources and constant leadership and supervision, workers are empowered to self-manage the processes and activities necessary for their jobs independently or in conjunction with their immediate management team. New work systems are adaptive, flexible, and provide better work-life balance structure for the new workers. Organizations inevitably need to modernize their work systems and their technologies to provide self-managing work systems for the new workers of the future.

Organizations should create work structure and systems that empower their employees' work self-management. They should be more outcome-driven, provide easy access to necessary informational resources for projects and work responsibilities, and be mindful of their expectations from the employees who work remotely. New works should be evaluated by a management team that is aware of the nature of the new works and their flexible hours and workflows. Overwhelming demands and noncommunicative rigid governance make new works counter-effective. Financial and social incentives such as result-driven bonuses and in-person socializing gatherings help organizations to keep their new workers engaged and productive. New data workers should practice self-management in the absence of direct and constant leadership by learning non-routine cognitive skills to be successful in performing their tasks in the fast-changing technologically-enabled work environments. These skills require problem-solving, abstract reasoning, systems thinking, and critical and analytical thinking. New self-managing work systems are inevitable and organizations and their employees should be proactive and prepared to stay competitive and relevant in the new world.

REFERENCES

Alter, S. (2013). Work system theory: Overview of core concepts, extensions, and challenges for the future. *Journal of the Association for Information Systems, 14*(2), 72–121.

Avogaro, M. (2019). The highest skilled workers of Industry 4.0: New forms of work organization for new professions. A comparative study. *E-Journal of International and Comparative Labour Studies.* http://ejcls.adapt.it/index.php/ejcls_adapt/article/view/648

Brian, C. (2022). *Airbnb News.* Retrieved 3 July 2022, from news.airbnb.com/airbnbs-design-to-live-and-work-anywhere/

Cruise. (2022). Retrieved 3 July 2022, from www.getcruise.com/

Dadgar, M., & Joshi, K. D. (2018). The role of information and communication technology in self-management of chronic diseases: An empirical investigation through value sensitive design. *Journal of the Association for Information Systems (JAIS), 19*(2), 86–112.

HyFlex Learning Community. (2022). Retrieved 3 July 2022, from www.hyflexlearning.org/

Lund, S., Madgavkar, A., Manyika, J., Smit, S., Ellingrud, K., & Robinson, O. (2021, February 18). *The future of work after COVID-19 | McKinsey.* www.mckinsey.com/featured-insights/future-of-work/the-future-of-work-after-covid-19

Manz, C. C., & Sims, H. P. (1980). Self-management as a substitute for leadership: A social learning theory perspective. *Academy of Management Review, 5*(3), 361–368. Business Source Complete.

Zoom Video Communications, Inc. (2022). *Video Conferencing, Cloud Phone, Webinars, Chat, Virtual Events | Zoom.* Retrieved 3 July 2022, from https://zoom.us

Index

Note: Page numbers in **bold** refer to figures, and those in *italics* refer to tables.

Printed in the United States
by Baker & Taylor Publisher Services